Reaching All Students with Mathematics

Reaching All Students with Mathematics

edited by

Gilbert Cuevas
University of Miami
Miami, Florida

and

Mark Driscoll
Education Development Center
Newton, Massachusetts

NATIONAL COUNCIL OF TEACHERS OF MATHEMATICS

Library of Congress Cataloging-in-Publication Data

Reaching all students with mathematics/edited by Gilbert Cuevas
and Mark Driscoll.
p. cm.
Includes bibliographical references.
ISBN 0-87353-357-7
1. Mathematics—Study and teaching—United States. I. Cuevas,
Gilbert J. II. Driscoll, Mark J.
QA13.R43 1993 92-44679
510′.71′273—dc20 CIP

Printed in the United States of America

Contents

Preface

As recommendations for sweeping change in mathematics curriculum and instruction, NCTM's *Curriculum and Evaluation Standards for School Mathematics* and *Professional Standards for Teaching Mathematics* reflect a set of deeply held beliefs about students and learning, about teachers and teaching, and about mathematics. This book was born out of a conviction that one of those beliefs—that all students can learn a significant core of high-quality mathematics—should receive special attention because it is essential to the future of mathematics education and because it challenges some older, more traditional beliefs in the school community.

The book also arose from the conviction that the task of reaching all students is difficult, requiring extraordinary efforts and not merely fine-tuning of existing efforts. We must learn how to capitalize fully on diversity as a strength, not a liability, in the mathematics classroom and how to bring into the mathematics learning community all those students, with their diverse backgrounds and needs, who have been underrepresented in that community. Such sweeping change can occur only through extraordinary efforts.

NCTM provided the opportunity to put this book together when it convened us as a task force several years ago. Since each of us on the task force knew colleagues who were struggling to show that all students can indeed be reached with mathematics, we designed the book as an opportunity for those colleagues and others to tell their stories. Consequently, we invited manuscripts from all levels of precollege mathematics education that would portray programs designed to reach all students. We collected a set of draft manuscripts, and with the help of several teacher-reviewers (for whose efforts we are extremely grateful), we were able to shape the book you are reading.

The range of stories submitted was quite wide, and we chose to reflect that range in the book. Though it is our belief that reaching all students will, in the long run, require programs that transform old paradigms of teaching and learning, we became aware that few such programs currently exist. In addition to the descriptions of more radical departures from the status quo, we decided that the mathematics education community would also be well served by the accounts of educators who are increasing the numbers of students reached in mathematics simply by enriching and extending current efforts. Thus, in the resulting set of stories, you will see descriptions of several programs that move only incrementally away from traditional approaches. You will also read accounts of daring and revolutionary efforts to widen the participation and success of students from underrepresented groups—some through the invention of curriculum and instruction in line with the NCTM *Standards* and others through the invention of new approaches to the professional development of teachers.

We hope that you will find points of guidance in each of the chapters and will be inspired to undertake similar efforts to reach all students with mathematics.

NCTM Task Force on Reaching All Students with Mathematics

Gilbert Cuevas, Cochair
Mark Driscoll, Cochair
Noemi Rodriguez-Lopez
William Tate
Tonya Urbatsch

Part 1
Global View

Changing demographics have raised the stakes for all Americans. Never before have we been forced to provide true equality in opportunity to learn.

Everybody Counts (National Research Council.
Mathematical Sciences Education Board. Washington, D.C.:
National Academy Press, 1989, p. 1)

This statement summarizes the challenge facing mathematics educators in the remaining years of this century and the beginning of the next. The authors in "Part 1: A Global View" present the obstacles and factors we must consider if we are to meet this challenge successfully. Usiskin's chapter gives an overview of the difficulties students encounter in learning mathematics. He emphasizes the need for integrated efforts in curriculum and teaching reform. Looking at the learning of mathematics in a different way is the underlying message.

To meet the challenge we must also include all the "players" in the mathematics education game. Bezuk et al. describe successful strategies for involving the different "communities" that affect the learning of children. These include parents, college students, community agencies, and teachers working together to give students "true equality in opportunity to learn."

The goal of mathematics education for all must also deal with students' cultural diversity. Zavlasky gives an overview of the cultural factors teachers need to address. She also describes strategies for incorporating the students' culture into the mathematics-classroom environment.

1

Introduction
Reaching All Students: A Vision of Learning Mathematics

Lee V. Stiff

The National Council of Teachers of Mathematics (NCTM) has "the comprehensive mathematics education of every child as its most compelling goal." Indeed, the Council's creation of the *Curriculum and Evaluation Standards for School Mathematics* and the *Professional Standards for Teaching Mathematics* provides a framework on which teachers and students, parents and school boards, and community agencies and business can build strong mathematics programs and promote the full participation of all students. If you have not read them, consider these documents companion volumes to this book. Together, these works present a vision of how each student can realize his or her full potential. With this book, NCTM continues the discourse about ways to enhance mathematics instruction and identifies approaches that engage every child; thus the goal of reaching all students is closer to becoming a reality.

NCTM's position is clear. By *every child* is meant—

1. students who have been denied access in any way to educational opportunities as well as those who have not;
2. students who are African-American, Hispanic, American Indian, and other minorities as well as those who are considered to be part of the majority;
3. students who are female as well as those who are male;
4. students who have not been successful as well as those who have been successful in school and in mathematics.

The contributors to this book recognize that strategies and approaches for enhancing mathematics instruction must be comprehensive and flexible so that no student is left out.

There is a lot of interesting mathematics that many students never experience. The *Curriculum and Evaluation Standards for School Mathematics* identifies mathematical skills and concepts that should receive increased emphasis in mathematics classrooms because of their importance to real-life uses of mathematics. These include estimation, statistics, probability, and measurement. A more relevant and comprehensive curriculum offers greater opportunities for all students to engage in meaningful mathematics. Many of the skills and concepts that should receive greater emphasis can help teachers and students make mathematical connections and become an effective way to motivate students to study mathematics.

A comprehensive approach to mathematics instruction uses technology where appropriate and focuses on skills and concepts that are relevant to students now! Strengthening problem-solving and reasoning skills empowers students to handle real-life problems, which promotes a better understanding of mathematical skills and concepts. In this context, communication becomes an effective way to teach and learn mathematics and is the basis for developing meaningful interpersonal communications both in and out of the classroom.

The nature and quality of interactions in the mathematics classroom is a telling index of how well the needs of different students are being met. Some students will prefer the support of a group, others will want to work alone; many will prefer alternative teaching strategies and multiple representations of concepts. As students begin to see that a richness of points of view and approaches are valued, they will express their mathematical abilities in the ways most suitable to their experiences and personalities. As the *Professional Standards for Teaching Mathematics* suggests, teachers must be open to a variety of student expression and skillfully orchestrate that expression. Teachers' commitment to each and every child must be real and must go beyond a verbal pledge—their commitment to educate all students must be embodied in their actions.

As you read the chapters, reflect on why many students do not seek the study of mathematics and why many students have decided that mathematics is not relevant to their lives. Many of the answers to these and similar concerns can be found here. However, it should be acknowledged that a student's disposition to do mathematics depends, in part, on a teacher's disposition toward that student. As Usiskin points out in his chapter, opportunities to learn meaningful mathematics are not always the same for all students. Differentiated curricula and elitist notions about who should continue the study of mathematics have contributed to a general dislike of mathematics by many students, especially among girls and many nonwhite students. The *Curriculum and Evaluation Standards for School Mathematics* and the *Professional Standards for Teaching Mathematics* clearly indicate that all students can learn mathematics and must be given the opportunity to do so. Giving the opportunity to learn mathematics means that teachers start where students are, not where "they ought to be!" And from that starting point, teachers help students go as far as possible. (See chapters 6 and 13.)

The roles that teachers and students will play in tomorrow's mathematics classes will be different from those in teacher-centered classrooms. As mathematics teachers become facilitators of knowledge, students will assume a greater responsibility for their own learning. Do not be mistaken. Students are quite capable of assuming an expanding role in the mathematics classroom. Students have always made conjectures about mathematical skills and concepts (just witness the algorithms used by students to add fractions!), but now their attempts to make sense of their mathematical encounters should be focused and directed, with the help of teachers, so that students themselves can construct appropriate relationships among mathematical skills and concepts. (See chapters 5, 7, 10, 12, and 15.)

An important factor in students' assuming a greater role in the teaching and learning of mathematics is communications in mathematics classrooms. Chapters 4, 8, 9, 14, and 15 address issues related to mathematical communication. The importance that teachers place on student discourse is connected to the quality of student-teacher interpersonal communications. The quality of student-teacher interpersonal communications depends, in part, on issues of cultural diversity and a healthy respect for differences. (Chapters 2, 3, 8, and 13 address a different aspect of this issue.)

The students who have been the most successful in learning mathematics will also benefit from the teaching and learning strategies and approaches presented in this book. The mathematical content and its connections to real-life situations as well as to other mathematics; the mode by which instruction is delivered (using manipulatives and technology, for example); and a variety of assessment techniques that should be employed (journals, group assessment, or observations) reflect a broader acceptance of what constitutes evidence of mathematical accomplishment. Consequently, students who currently succeed in mathematics will simply have additional conduits by which skills, concepts, and relationships can be acquired and represented. But equally important, students who have not been successful before will have alternative paths to success.

NCTM has formed the Committee for a Comprehensive Mathematics Education of Every Child (SEE ME) to assist in implementing its stated priority of the mathematics education of every child. One of its goals is to "disseminate information about teaching practices and student learning related to the mathematics education of every child." SEE ME applauds the contribution of this book in meeting this NCTM goal. As SEE ME strives to affect practices and implement activities directed toward the comprehensive mathematics education of every child, it welcomes the efforts of all concerned.

The goal of every teacher should be to provide all students with the "opportunity to develop their full human potential in an environment where cultural differences are respected and valued and where full participation and partnership are the norm." Cooperation among students, teachers, and parents to achieve student success (see chapters 2, 10, 11, and 12) tends to create an empowering learning environment in which students are expected to achieve, and do. The vision of learning mathematics presented by the collected works of this book is already realized by many students, teachers, and parents. You now have the opportunity to share in that vision.

2

If Everybody Counts, Why Do So Few Survive?

Zalman Usiskin

Since 1983, the scope of the University of Chicago School Mathematics Project (UCSMP) has been such that we have dealt with all the grades from kindergarten to grade 12, with models for curriculum and teacher training, and with materials for both students and teachers. The complexity of our enterprise mirrors the complexity of the task of school mathematics, and in attempting to improve school mathematics, we have had to deal with virtually all the problems associated with it. The subject of this paper is as difficult a problem as any in school mathematics. Perhaps for that reason, there has been until recently very little discussion of it. The problem is the fact that only a small percentage of our students survive their school mathematics experience.

It is impossible to determine exactly what the percentage of survivors is, because the percentage depends on one's definition of "survival" and because mathematics courses through the United States and Canada are not uniform. For instance, NCTM has recommended since its *Agenda for Action* (1980) that three years of high school mathematics be required of all students. Suppose we take "survival" to mean that a student completes the equivalent of two years of algebra and a year of geometry, the most common program. Then a self-reported statistic by the seventeen-year-olds in the 1986 National Assessment would indicate that nearly half our students survive, since about 47 percent of seventeen-year-olds reported a half-year of algebra II or precalculus or calculus as their "highest level taken" (Dossey et al. 1988, pp. 116–17). However, algebra II in many schools is the title of the second semester of first-year algebra. Also, some students take second-year algebra before geometry, and so the "highest level taken" is somewhat misleading.

This chapter is adapted from a talk given at a UCSMP conference in November 1989.

Furthermore, since some students have already dropped out of school at this age, NAEP data are likely to be optimistic. With all this, I would estimate that the percent of survivors under this first definition—those who take three years of secondary school mathematics—in 1986 was no more than 40 percent.

Regardless of the accuracy of the figure, the number of survivors is increasing. The same item has been on the last four National Assessments, and there has been about a 5 percent increase in the number of seventeen-year-olds reporting having taken either second-year algebra or precalculus or calculus. In the 1990 Texas state adoption, the number of second-year algebra texts purchased (about 120 000 in a population of about 17 000 000) would indicate a survival rate in that state (perhaps typical of the entire U.S.) of from 40 to 50 percent. Furthermore, according to ACT data the survival rate is as high for females as for males. The message that more mathematics is needed is getting around. Under this definition of survival, the cup is almost half full.

From recent Bureau of Labor Statistics (1990) and unemployment data, using the age cohort of the class that graduated from high school in 1988, best estimates are that 48 percent of the population of graduating seniors now go to college. It seems, therefore, that we can equate survival in school mathematics with college entrance. This agrees with remarks by Donald Stewart, president of the College Board, "The link between math and college is 'almost magical'; math is the gatekeeper for success in college."[1]

If the cup is almost half full, then it is a little more than half empty. Among those who do not take three years of high school mathematics are huge numbers of students who have had no algebra or just a year of algebra, no geometry, no statistics, no probability. A third of this population enters the job market either before graduating from high school or immediately after graduation. A further sixth of this population becomes unemployed. These youths, far more likely to be from our cities, are simply unprepared for the mathematics they will encounter on almost any job. It is not that they have not learned what they were taught; these students never had the content.

When we compare the situation to that in other countries, we realize how empty the cup is. In many countries of the world, the equivalent of two years of algebra and a year of geometry are completed by the end of tenth grade, when almost all students are still in school and when almost all students are taking the same, required mathematics.

The National Assessment data are the most optimistic data available regarding survival. Other statistics show that there is quite a drop-off after second-year algebra. Only about 12 percent of the age cohort in the United States in 1981–82 was enrolled in precalculus or calculus classes as seniors in

---

1. As quoted in "Math Key to Success in College," by Pat Ordovensky, in *USA Today*, 4 October 1990.

high school. This is another estimate of our survival rate. That 12 percent is about the same survival rate as in Japan and many other industrialized countries. However, the Second International Mathematics Study (commonly known as SIMS) reported: "In most countries, all advanced mathematics students take calculus. In the U.S. only about one-fifth do." Kifer, one of those who worked on SIMS, has commented that we sort so early that by grade 8 the proportion of students taking algebra is about the same as that of those taking senior-level mathematics in grade 12 in other countries (Kifer 1986, 1989). An extensive look at this problem has been done by the sociologist Elizabeth Useem (1990).

If we make a standard of performance part of the definition of survival, then the situation is even more bleak. Even our best survivors do not perform so well. Here is what SIMS had to say about the performance of our seniors: "The achievement of the Calculus classes, the nation's *best* mathematics students, was at or near the *average* achievement of the advanced secondary school mathematics students in other countries .... The achievement of the U.S. Precalculus students (the majority of twelfth grade college-preparatory mathematics students) was substantially below the international average" (McKnight et al.1978, p. vii).

The highest survival rate among the twenty-three countries that were involved in SIMS was in Hungary, where 50 percent of the age cohort took the standard mathematics curriculum as seniors. British Columbia (a "country" for the purposes of SIMS) was second, with 30 percent, 2.5 times the U.S. survival rate. Even with the higher survival rate, students in both of these geographic entities outperformed students in the United States. (It is not clear how the rest of Canada would have performed; only Ontario and British Columbia were represented in the study.) The population of British Columbia is about the same as the population of Chicago, a little less than 3 million, but in Chicago less than 50 percent of the age cohort graduates from high school, and the percentage who have taken senior mathematics is probably less than 10 percent. The survival rate in our inner cities is comparable to the survival rate in third-world countries.

Why do we lose so many students? Of course some of the reasons are societal. Drugs are a scourge, some parents do not value education and discourage their children from giving their all, and many students do not have guidance from home. But this does not account for most of the mathematics dropouts. Even in our best schools, even from the best homes, there are children who need to make it and don't. If "everybody counts," as the National Research Council report of 1989 was aptly titled, we must ask why so few survive.

A Dangerous Myth

There are a number of myths that one must dispel if one is to make the kinds of changes we need to make to update and upgrade our school curriculum. One of those myths is fundamental to the problem of survival: *Either you have it in math or you don't, and the job of the teacher (or school) is to find out who has it.*

Everyone knows some teachers who believe the first part of the statement, and we all probably know some teachers who act in accordance with the second part of the statement. This essay does not focus on those teachers, because I believe teachers are not the major cause. I believe the problem is more subtle than this. At the same time that we have people saying that everyone can learn mathematics, we have a system in which certain common practices serve to undermine that belief.

The difficulty of the issue is that a statement very similar to the myth is true: Some people can learn more mathematics and do know more mathematics than others. Should we treat people differently but offer the same opportunity, or should we treat everyone the same? We desire equality or equity in a democracy, but there are two common interpretations of equality: at times equality means equal opportunity; at other times equality means equal treatment.

A Story from Elementary School

Generally in education our rhetoric is that we give all students equal treatment, but our practices show that we prefer equal opportunity. The problem is that the opportunities are cut off at very early grades. I would like to give a personal example of this. In telling this to other groups, the response has been that my story is not unusual and, in fact, is nowhere near a worst-case scenario.

My son, who is now in sixth grade, took his first college entrance exam at the end of first grade. (I used to report that this occurred at the beginning of second grade, but I found out recently that it was earlier.) By this, I mean that he took an exam where the results will affect what colleges he might attend. The exam was not something that his mother and I pushed him to take; it was given by the school. I have been told that it included reading, and the mathematics covered addition and subtraction facts up to 18 and some names of geometric figures. That was it.

He knew everything on the exam because my wife and I are educators and had taught all that stuff to him before he even started first grade; as a result of his performance, he was placed in a special second-grade math class. (There was no special class for any other subject in the school.) That math class reviewed the first-grade curriculum quite a bit even though the students already knew first-grade mathematics, but it did not spend as much time reviewing as the other second-grade classes. This enabled my son's class to cover more material than the other second-grade math classes.

The school district we were in did not track. So, in third grade, the procedure was repeated. Students were tested, and those who scored highest were put in a special third-grade math class. Now who was most likely to score highest? Those who were in the second-grade special class, of course, because they had done more the previous year. Each year the district did the same thing, and each year the students that have been in these special classes had a greater and greater chance to remain in the top class. You can see that there is no tracking per se, only equal opportunity each year. When seventh grade comes, this school system has an algebra class, and all the students like my son will have a high probability of being in it; if they continue they will take Advanced Placement calculus in eleventh grade. In twelfth grade they will be able to take other such advanced courses, and this will help their chances of getting into one of the better colleges. In this way, that test at the end of first grade will have influenced the colleges they will get into, and thus it seems appropriate to call the exam my son's first college entrance test.

What about those second-grade students who did not know their facts to 18 at the end of first grade? Like me, their parents were not told of the significance of the test; they were not even told there would be a test. Each year that a student is not in the best class, that student falls farther and farther behind the best students. But there is something even more significant: As far as I could tell, and I was monitoring the work all year, in both third and fourth grade there was nothing the special math class did that could not have been done in the standard classes. The special math class reviewed previous years less, which is wise policy for virtually all students. They did open-ended problems and "word problems," which is wise for all. The students had discussions about mathematics in their class, which is wise. In short, these students got the kind of education all students should have had the opportunity to have.

Obviously there is a need to do something for students in a grade who know all or virtually all the mathematics to be discussed at that grade. Something more should be done with them. It is boring and self-destructive to have them learn the same mathematics as other students who have not progressed as far. But, as I have tried to argue, it is self-destructive to the others to withhold reasonable opportunities from them.

There is a happy ending. The story you have heard is true, and I first told it in November 1989 at a UCSMP conference. There were people from my son's school district in attendance, and my remarks concerned them. The part of the story that got them thinking most dealt with the quality of the curriculum. The math committee got together and decided that in fact they were depriving all their students. Last summer they decided to upgrade the curriculum for the other students so that it would resemble the curriculum they have been giving to the best. This meant that they accelerated *every* student in those classes an entire year, and they did this with no advance warning. They simply gave students in grades 3–6 books that were meant for grades 4–7. The verdict so far is that other than a problem at the beginning of grade 6 (the grade 7 books in this basal series had never been used by the district, and the jump was more than they expected) the entire venture has been an outstanding success.

It is possible to do this in almost any school or school district. As Fuson, Stigler, and Bartsch (1988) have reported, we teach even the simplest of mathematics content later than many other countries. When SIMS entitled one of its reports *The Underachieving Curriculum,* it was partially in recognition of the fact that the United States seems to have the weakest K–6 curriculum of any industrialized country.

Middle and Junior High School Practices

When we move into middle and junior high school, a different phenomenon begins: the beginning of remedial classes. We take those who are behind and slow everything down, so they get even further behind. We implicitly tell them they are dumb by not giving them anything of interest, often not giving them anything new, but spending the entire year reviewing things they have had before. The goal of remediation is to save these students, to bring them back into the mainstream, but how many come back? Very few. As far as I know, we are the only country in the world that takes students as young as twelve years old and sorts them out of the mainstream.

The problem is complicated: it is not the grouping that causes the problem but what we do in the groups. The mathematics of this is simple: If student A is behind student B and student A is being taught at a rate x that is slower than the rate y at which student B is being taught, then student A will not only remain behind student B but become further and further behind.

We must change our rhetoric to students who are in remedial programs. We must tell them, "Mathematics is important. It is on every test you might take to get a job in business or industry. You need it as a consumer. It is part of literacy. It is 50 percent of the SATs. But you are behind. *You must work harder because you are behind, and we will help you do so.*"

Senior High School Practices

In high schools, the tracking issue is overt. There are levels of algebra in almost every school of sufficient size. If you are at the low level of algebra, you can get into the low level of the next course. If you are at the middle level, you can get into the middle level, unless you do not perform well—then you go into the low level. If you are at the high level, you can stay at that level, or if you do not succeed, you can go down a level. From one year to the next, you can almost never go up a level, because the courses are designed to widen the gap in performance. If you have *just one bad year,* you are dropped from the highest level and cannot return.

We insure that some of the best students go down levels because, in the name of high standards, we give them wipe-out courses. In 1981–82 about 13 percent of the students in the country took algebra in eighth grade (McKnight et al. 1987). Since about 7 percent took precalculus or calculus as juniors in 1986 (Dossey et al. 1988), we wiped out about half of these students—our very best students—in three years. More will choose not to take calculus as seniors, and I estimate that only about a third of all students who are identified in seventh grade as capable of advanced placement actually become advanced placement students. The practice is ingrained: it is the expectation in many school districts that of the two or three classes of eighth-grade algebra offered, there will be only one class of survivors who reach calculus.

What seems to cause the wipe-out are two practices that we must work to remove. The first is to expect all these students to be future mathematicians or mathletes. We teach them theory without any application; we give them content that is not useful even for future mathematicians but is found on math contests. We look at books for such students and say that book is OK but not hard enough. That is, we consider difficulty—mere difficulty—to be a virtue even more important than the content of the course.

We have direct evidence of this belief. We know of people who look at the UCSMP courses at the secondary level and say, "They have nice stuff, but they are too easy for our best students." My response is a question: Which is more important, that students be exposed to the right stuff or that the course be difficult? In many schools, the first objective in selecting the curriculum for the best students is that the course be hard. If ever there was an example of mixed-up priorities, it is this.

The second practice causing the wipeout is the tendency to have moving standards. This phenomenon occurs far more widely than in classes for the best students. Regardless of how well students perform, if a teacher were to give them all As or Bs, in many schools that teacher's job would be in jeopardy. The teacher grades too easy. Success is not expected by all, particularly in mathematics. Again it traces back to the myth that you either have it or you don't. And there are levels of having it—Becky is a B student, Charles a C student. What is wrong with giving all students As?

It is useful to realize that five hundred years ago the arithmetic we teach today in fourth grade, namely partial product multiplication and long division, was a college subject (Swetz 1987). It was unfamiliar and new and therefore thought to be difficult. That is no longer the case. Mathematics is easier today. We have better algorithms, among them the punching of keys on a calculator for computation and for graphing.

The Influence of Tests

Curves are related to another of the pernicious practices, the normed test that gives grade-level equivalents or percentiles. The problem here is that someone has to fall below the 50th percentile; in fact, half the population has to fall there. And this tells you nothing about how well they are performing. You would think that mathematics teachers would understand these relative standards better than others, but I have heard many teachers identify children as being in a particular stanine as if this is a tattoo that cannot be removed. "We put 2d–4th stanine children in that class, 5th–8th in the other class, . . ." It is that myth again: not only do some people think they can determine who has it and who does not, but they act as if they can determine precisely the extent to which every student has it.

Normed tests give rise to some interesting rhetoric: the rhetoric of the overachiever. The child performed well but surprised us; according to the test the child was not supposed to be able to do that. To me there are no overachievers, there are just tests that were incorrectly employed to predict that someone could not do something.

How Many Are Really Disabled?

Allow me another true story that is too familiar to many parents. In the early days of UCSMP, I was writing and teaching the first draft of one of our courses. Every day for the school year, I taught a class of ninth graders who did not get into algebra, the 10th to the 25th local percentile in this school. Since I was teaching only this one class, and even that class was legally assigned to a faculty member of the school, I was not aware of all the rituals of the school, and in November I received a computer list of my class on which there were asterisks (*) by four students' names. I was puzzled. I went to the department chair and asked what the asterisks meant, and he responded that those students were LD (learning disabled). I responded that I had been grading the papers of these four students every day since the beginning of the year and I had no evidence that they were learning disabled. I asked rhetorically, Were they cured?

There are students who are truly learning disabled, but it also happens that we in the United States seem to have the greatest percentage of learning disabled students in the world. These students that I was teaching were not identified as LD in ninth grade; in this school district they are identified as early as first grade! Can you imagine going through school with an asterisk by your name that announces to the world of teachers that you have a learning problem? How many students are learning disabled because they are doing what is expected of them? One of the reasons that so few survive is that we write off students, often very early.

Virtually anyone could do a little better than he or she is doing, so in that regard almost any student is an underachiever. When a student comes up to me and says, "You know, Mr. U, I could have gotten an A on that test if I had studied," my answer is simply, "I don't have any evidence of that, and neither do you. But I think you're right, except you may have to study more than you thought." When a student comes up to me and says, "My parents couldn't do that homework either," I would respond, "Five hundred years ago [in 1489] the symbols for plus and minus were first used. If every generation knew only what its parents knew, you wouldn't even know how to check whether you are being charged the right amount in a store. No one would know how to drive a car. Every generation needs to learn more than its parents know; otherwise, there would be no progress."

Curriculum as a Cause

The curriculum is not only underachieving, it is often unappealing, and this constitutes still another cause of the dropout rate. The curriculum we have today is, for the most part, the curriculum of fifty years ago with changes caused by a combination of two movements, new math and back to basics. On the good side, new math gave us the first impetus for a broad-based curriculum in the elementary school, a mathematics curriculum rather than an arithmetic curriculum. At the high school level, new math also caused some shifting of areas, combining plane and solid geometry, moving trigonometry in with algebra, integrating analytic geometry with other high school mathematics. But new math assumed that all students were motivated like potential math majors, that is, it assumed students did not need motivation, and the texts of that era omitted virtually all the real-world connections that had been in the curriculum.

The era of new math lasted from about 1959 to 1972. The reaction to new math, back to basics, carried through the early 1980s, assumed that all students were incapable of understanding any mathematics, and incapable of applying any mathematics and gave us a curriculum even more devoid of motivation. The lack of appeal in the curriculum—day after day of drill sheets and repetitive activities in the elementary and junior high school, algebra taught as a foreign language, geometry taught without connections to any objects in the physical world—is certainly a major cause of the high attrition rate in school mathematics. Why keep studying a subject beyond arithmetic when the only reason given for studying it is "you need it for the next course"?

Although the problem-solving movement of the 1980s was clearly conceived to counter the back-to-basics movement, the problem-solving movement can also be seen as an attempt to bring some appeal, some motivation back into the curriculum. This movement had some impact on elementary school books but almost no impact at the high school level. It never picked up steam because of the vagueness of the idea of problem solving. What is done in one grade is almost never followed up in the next grade. Thus there are no long-range effects.

It is significant that NCTM got off its singular devotion to problem solving with its *Curriculum and Evaluation Standards* (1989). The work of UCSMP, of COMAP, and of many other groups has already had its effects. The mathematics curricula being recommended by virtually every group in the country are far more appealing than any we have seen in the past thirty years. It is a good time for mathematics education.

Some Solutions

Having identified problems at all the grade levels, let me now speak of the solutions from the perspective of our project. Some of the practices I have been encouraging or discouraging are not in the domain of UCSMP work; one could do them or undo them with any curriculum. But some solutions are intimately related with our work. At this point allow me to reiterate that we are not the only people working for solutions, and fortunately for all of us many of the solutions are being promoted by NCTM and by the Mathematical Sciences Education Board.

1. *Start from where students are.* This axiom of good curriculum and instruction has three corollaries. (*a*) *Do not needlessly review.* UCSMP is known for our views on this subject (Flanders 1987). Our primary school curriculum takes advantage of what students know to go further than any published curriculum at kindergarten and first grade. Our secondary school curriculum avoids the counterproductive time spent on review that is inflicted on most students in grades 7 and 8.

Do not interpret this advice as meaning that you should never review. We believe strongly in reviewing, but in reviewing only what needs reviewing and, whenever possible, in the context of learning new material.

(*b*) *Do not destroy students by giving them courses for which they are unprepared.* Although we are careful to indicate that the first course in our secondary school curriculum requires that students be at the seventh-grade level in mathematics, we see some school districts using the materials with students who are not that well prepared. We see states like Mississippi and Louisiana and cities like Chicago requiring all students to take algebra without changing the curriculum sufficiently before algebra to enable them to succeed. The first year of the required algebra course in Louisiana, fewer than half the students completed the course. Equal opportunity is hollow when the failure rate in a course is more than 50 percent.

Many of you have seen the film *Stand and Deliver.* The success at Garfield High School in Los Angeles is real; it is possible to take students who are far behind and bring them up even to the level of top Advanced Placement calculus students. But the contract these students sign requires them to spend thirty hours a week on mathematics. Simply put, they need to make up for all the lost years. *You are behind and so you must work harder to catch up.* There is no quick fix to our problems in school mathematics; to change student performance significantly requires a significant change in what students encounter.

(*c*) *Allow students of different ages to do the same mathematics.* There will always be students who are quite ahead of their peers, either because their parents have taught them or because they are interested and read on the side or because they are in some special programs. It is senseless to keep someone at a particular level in any subject because "if I teach this, what will next year's teacher do?" We would not tolerate keeping a piano student on a particular piece even though the student has learned it. We would not tolerate having a child read the same story over and over. We keep them interested by moving on. Part of the reason the dropout rate is so high is that the school mathematics curriculum is so boring for so many students, both the poor and the good. The boredom starts in elementary school, but it continues through high school and even college.

Our evaluation component has completed a number of studies at grades 7–10 with students at different grade levels who come in with the same knowledge. Equating for entering knowledge, the performance of these students is independent of age. We have yet to find any evidence of that thing some people call "mathematical maturity," which is supposed to be obtained at some particular age.

2. *Provide remediation immediately and powerfully.* Students do differ. Unless a teacher takes pains to keep them at the same level, there will be students who are unable to proceed at the same pace as most of their peers. Remediation must be immediate. Do not allow students to fall behind. For instance, in October, in most classes there are students who are not learning what is being taught. Teachers often think, "If that student does not perk up, next year that student will be in real trouble." That thinking is too lackadaisical. If a teacher is content to wait until next year, the student is content to wait also, and the importance of learning the idea this year is diminished. Students need to shape up as soon as the deficiency is noted. They need after-school or before-school help with their work. Many teachers use the quicker students to help the slower students.

If students don't shape up during the school year, they need summer programs to get them back with their peers. The Algebra Transition Project in Philadelphia is an example that this approach can work. School boards and school administrators generally understand that mathematics is important, a key to success; acting on that understanding requires that programs be instituted to ensure that as many students as possible succeed. It is best to start such programs in the early grades, when the differences among students are the narrowest.

Even with these suggestions—powerful remediation for the students who are behind and moving students up who are ahead—classes will necessarily have a wide range of students in them. There are two aspects of current reform that help reach such a wide range. The first is technology.

3. *Use technology.* Technology is the great equalizer. For instance, research indicates that calculators seem to diminish differences in performance between girls and boys. We have had evidence from teachers for many years that calculators help to diminish differences between better and poorer students, we believe because even poor students can think and use strategies. Often they became identified as poor students only because they were not so good at paper-and-pencil arithmetic.

Ironically, technology is also considered by some people to be a problem in terms of equity. Here I believe strongly that calculators must be treated in a different way from computers, because they are so cheap and so widespread. On the 1986 National Assessment, 82 percent of nine-year olds, 94 percent of thirteen-year-olds, and 97 percent of seventeen-year-olds reported that there is a calculator at home (Dossey et al, 1988). The cost of a single personal computer and a modicum of software, which schools seem very willing to spend, equals the cost of three or four classrooms of scientific calculators and a school's worth of four-function calculators. You do not get plaudits from your professional organization if you are against calculators in classrooms, but we know there are many people who are opposed to their use. Those who state that equity is a problem when it comes to calculators are either misinformed or purposely putting up a smoke screen to hide the fact that they are against calculators.

I am convinced that one of the reasons that calculators are not on some standardized tests is that they would make many of the traditional questions unusable, since they would make them too easy. It is that myth in another form: Mathematics has to be hard to be good. Still another way of putting the myth: If mathematics does not sort, then it is not real mathematics.

With respect to equity, the computer issue is different, since computers require significant expenditures and some districts can afford far more than others. Fortunately, however, computers are not as controversial as calculators. Perhaps some people do not realize that computers can do all the arithmetic calculators can. (Maybe I shouldn't have let the cat out of the bag.) Anyway, computers are glamorous, and we have found that PTAs and bake sales can earn good money for buying computers for classrooms even in places where the district cannot afford them. One computer in a mathematics classroom is essential. Here is what the NCTM *Curriculum and Evaluation*

Standards says about this (1989, p. 8): "Because technology is changing mathematics and its uses, we believe that appropriate calculators should be available to all students at all times; a computer should be available in every classroom for demonstration purposes; every student should have access to a computer for individual and group work; students should learn to use the computer as a tool for processing information and performing calculations to investigate and solve problems."

The fundamental difference between what is said in the *Standards* and what has been said in many previous reports is that the most important use for the computer is as a tool, not as a teacher. There is no mention of CAI in the old sense of the computer as a tutor. The role of the teacher is at least as important with computers as without them. Most leaders today have come to realize that the best use of the computer in classrooms is to use it there as it is used in the world outside the classroom—as an extraordinary storage device and tool and picture generator with a multitude of uses. With its ability to capture, inform, and motivate students, the computer is something we cannot afford to ignore. In general, the power of calculators and computers to do mathematics is what makes them an essential ingredient of a good mathematics program.

4. *Incorporate applications and real (not contrived) problem solving into the mainstream of the curriculum.* The second feature that makes work with a wider range of students possible is the use of problems that relate to a lot of different ideas, problems rich with context. If these types of problems are incorporated into the curriculum, then the students who have grasped the easier ideas can focus on the harder ideas. Everyone who has done this notes that students who cannot compute can often do everything else. So make certain there are calculators to help them.

Having a context helps more than just dealing with the wide variety of students. Imagine a reading curriculum in which day after day students learn new words without knowing what they mean and read sentences, but not in stories. Furthermore, imagine a story in which there was no plot, in which Goldilocks and the Three Bears are on one page but four bears appear on the next and two on the third. That is the way that mathematics is taught to, and consequently assimilated by, many students. One week they get the addition of fractions, the next week it is subtraction; or one week they get addition and subtraction of polynomials, the next week it is multiplication, and the following, division. The rules come without reason: "In multiplying fractions, multiply numerators and denominators, but don't do that when adding fractions." "Add exponents when the bases are alike; subtract when dividing; multiply when it's a power of a power." No context; no way to check the problems; nothing to hang it on except memory. Mathematics is called abstract, but when taught this way it is not abstracting anything; it is gibberish, not much different from nonsense syllables.

Word problems are not automatic substitutes for applications. For instance, consider this word problem. There is a train. The train leaves a station one hour before a plane flying overhead in the opposite direction going three times the velocity that the train had when it was twice as old as the plane, which is three years younger than the station itself. The number of the train is a three-digit number; the tens digit is bigger than the units digit and the sum of the digits is 26. And on the train there is an engineer who himself is a third as old. There is a club car at the back of the train. In the club car they sell mixed nuts, some at $1.89 a pound. The engineer has a niece and a nephew on the train and sends the nephew back to the club car to get the mixed nuts. He gives the nephew 14 coins, some dimes and some quarters. It takes the nephew 20 minutes to get to the back of the train. The engineer is getting hungry and so sends his niece to the back of the train. It takes her 15 minutes. The question is, How long would it have taken them if they had gone back together?

As bizarre as that example is, it only caricatures what is found in most books. If there is anything that we have come to believe in more and more strongly at all levels as we work on UCSMP materials, it is the need for context. That context need not always be a real-world application; it can be a mathematical context—for instance, if multiplication is known, division can be related to it. It can also be a rich problem-solving context—for instance, the search for a function that has certain properties. But the real world supplies wonderful contexts for virtually all precollege mathematics, and one of the proudest things we can say about UCSMP materials at all levels is that we take advantage of the wonderful applicability of mathematics. We receive continual reports from (obviously surprised) teachers: "Not once this year has a student asked, 'Why are we learning this?' "

The movement toward applications, now one of the strongest movements in all of mathematics education, can be viewed as a refinement of the problem-solving movement. Do not give students only abstract problems; give them problems that have utility, if not directly for them, at least for someone in whose place they could imagine being.

Implementing technology and rich mathematics experiences on a wide scale in the elementary school is difficult if not impossible with the teaching corps we have today because so many of these people were sorted out of mathematics when they went through school. It is ominous that the one mathematics course required of prospective elementary school teachers is taught at a level no higher than an average first-year algebra class. One reason we in UCSMP believe that we must have specialist mathematics teachers in the upper elementary school (a belief also held by NCTM) is that we do not see how all teachers can be retrained to be able to give students the mathematics they need.

Summary of Problems and Solutions

The systemic problems causing the lack of survival can be summarized as follows: In many school districts, the better students are put into classes that enrich them with a curriculum that would be appropriate for all students, and the poorer students are put into remedial courses that cause them to lag further behind. There are many levels, and though few school systems have rigid tracks, it is very easy to go down a level but almost impossible to go up. For the best students, wipe-out courses, purposely made difficult, are created with the expectation that not all will survive.

These beliefs are supported by the treatment of normed tests as if they are indicators of absolute intelligence or ability to perform and by a guidance structure that identifies students as learning disabled but never seems to cure them.

We know that virtually all students can learn a significant amount of mathematics, since they do so in other countries and since programs in the United States and Canada have demonstrated that it is possible. Yet if the solutions were easy, these systemic problems would not be with us. Here are what seem to be the common threads of successful programs: They start from where students are. They do not needlessly review. They have high expectations, but they do not destroy students by placing them in classes for which they are unprepared. They allow students of different ages to do the same mathematics. As soon as deficiencies are found, powerful remediation begins. They use technology. Their curricula incorporate applications and problem solving. Perhaps most important, however, is a fundamental underlying belief quite different from the myth mentioned earlier: *Virtually everyone "has it" in mathematics, and the job of the teacher (or school) is to help the student realize that potential.*

Everybody Counts

I have tried here to deal with what is perhaps the major problem for the health of our country—the growing disparity between the haves and the have-nots, between those who are successful survivors of schooling and those who are not. Although my remarks discussed teachers and students, many of them could be extended to administrators and school districts.

One of the gnawing problems with trying to improve the mathematics education of average students, of the vast majority, is that we have greater difficulty reaching the school districts that need us most. One can only conclude, judging from performance on tests, that we have remedial school districts in our nation, not just remedial students. For example, in 1989 more than half of the schools in the city of Chicago, 25 of 47 schools in which more than forty students took the ACTs, scored in the bottom 1 percent of schools in the nation on those tests. The level of performance in Chicago and many other school districts is low on more variables than just student test scores; such school districts find it more difficult to send people to conferences, and hence their teachers, who often are not so well prepared in the first place, fall further and further behind.

Just as remedial students need to work harder, those school districts need to work harder, and they need more opportunities for their teachers. Projects like UCSMP cannot do it alone. I implore readers who are in higher-performing school districts to work with your neighboring districts. Show them what you do; show them what your students do; show them what technology you use and how you use it. Just as the health of a school depends on how all its students perform, the health of your community depends on how all its schools are performing. The good performance of the schools in our cities as well as in our suburbs, small towns, and rural areas is necessary for our collective health as a society. Everybody counts; let us work together to see that as many as possible measure up.

References

Bureau of Labor Statistics. *Report USDL 89-308.* Washington, D.C.: U.S. Department of Labor, 1990.

Dossey, John A., Ina V. S. Mullis, Mary M. Lindquist, and Donald L. Chambers. *The Mathematics Report Card: Are We Measuring Up?* Princeton, N.J.: Educational Testing Service, 1988.

Flanders, James. "How Much of the Content in Mathematics Textbooks Is New?" Arithmetic Teacher 35 (September 1987): 1823.

Fuson, Karen, James W. Stigler, and Karen Bartsch. "Grade Placement of Addition and Subtraction Topics in Japan, Mainland China, the Soviet Union, Taiwan, and the United States." *Journal for Research in Mathematics Education* 19 (November 1988): 449–56.

Kifer, Edward. "Opportunites, Talents, and Participation." In *Student Growth and Classroom Process in the Lower Secondary Schools,* edited by Leigh Burstein. Champaign, Ill.: Second International Mathematics Study, 1989.

________. "What Opportunities Are Available and Who Participates When Curriculum Is Differentiated?" Paper presented at the annual meeting of the American Educational Research Association, San Francisco, 1986.

McKnight, Curtis C., F. Joe Crosswhite, John A. Dossey, Edward Kifer, Jane O. Swafford, Kenneth J. Travers, and Thomas J. Cooney. *The Underachieving Curriculum.* Champaign, Ill.: Stipes Publishing Co., 1987.

National Council of Teachers of Mathematics. *An Agenda for Action.* Reston, Va.: The Council, 1980.

________. *Curriculum and Evaluation Standards for School Mathematics.* Reston, Va.: The Council, 1989.

National Research Council. *Everybody Counts: A Report to the Nation on the Future of Mathematics Education.* Washington, D.C.: National Academy Press, 1989.

Swetz, Frank. *Capitalism and School Arithmetic: The New Math of the Fifteenth Century.* LaSalle, Ill.: Open Court, 1987.

Useem, Elizabeth L. "Getting on the Fast Track in Mathematics: School Organizational Influences on Math Track Assignment." Paper presented at the annual meeting of the American Educational Research Association, Boston, April 1990.

3

Educators and Parents Working Together to Help All Students Live Up to Their Dreams with Mathematics

Nadine S. Bezuk, Barbara E. Armstrong, Arthur L. Ellis, Frank A. Holmes, Larry K. Sowder

"Children will live up to our dreams, if only given the chance. Through this program, they can!"

This comment was made by one of the parents attending a parent-child workshop of the San Diego Mathematics Enrichment Project. It is representative of many of the remarks made by parents and teachers involved in this project.

The San Diego Mathematics Enrichment Project (SDMEP) is a multifaceted program aimed at enhancing the mathematics education of African-American and Hispanic second-, third-, and fourth-grade children in primarily low-income areas of San Diego, California. Focusing on students' mathematical experiences in school, after school, and at home, the SDMEP involves four groups: children, their parents, their teachers, and college students preparing to be elementary school teachers.

The project described herein was directed by Frank A. Holmes and supported in part by the National Science Foundation (SDMEP: San Diego Mathematics and science Academic Enrichment Project) and by the California Postsecondary Education Commission (MATES: Mathematics Accented for Teachers in Elementary schools). Opinions expressed are those of the authors and not necessarily those of the Foundation or the Commission.

The authors would like tot hank Bonnie Telfer, Heidi Janzen, and Antionette Lopez for their assistance in the collection and transcription of data described herein as well as their organizational support throughout the project. The authors would also like tot hank H. Vance Mills of the San Diego Unified School District for his assistance in establishing and supporting the ongoing efforts of the project.

Need for the Project

Why is there a need for intervention in the early years of elementary school education in mathematics and science for underrepresented ethnic students? Rothman (1988) reported that African-American and Hispanic students in Montgomery County Schools in Maryland had lagged behind their white and Asian counterparts in mathematics achievement by third grade. Similar trends appear in the Berkeley (California) Unified District (1983). According to Matthews (1984), many minority students do not understand how mathematics is used in everyday life and tend to think of mathematics as something done only with paper and pencil in the classroom. Although many minority parents want their children to do well in mathematics, they often do not know how to help their children with homework or with decisions regarding future educational plans.

The Quality Education for Minorities (QEM) Project (QEM 1990) set forth an action plan that contained guidelines and recommended strategies for improving the education of minority students. Some of these strategies included enhancing students' learning by extending "the school day and year to minimize summer loss and maximize exposure to mathematics and science" (p. 58) and strengthening teachers of minority students by paying them to work "12 months, and have them use that time to prepare for and to deliver quality education to students" (p. 70). The SDMEP has shown that many of the QEM recommendations can effectively be realized.

Many elementary teachers have limited understanding of mathematics, often accompanied by fear of the subject. This situation is discussed in *Everybody Counts* (National Research Council 1989): "Too often elementary teachers take only one course in mathematics, approaching it with trepidation and leaving it with relief. Such experiences leave many elementary teachers totally unprepared to inspire children with confidence in their own mathematical abilities" (p. 64).

It is especially critical that African-American and Hispanic teachers gain the confidence and knowledge necessary to reach students effectively. When their own knowledge of mathematics is broadened and their confidence in teaching mathematics is strengthened, African-American and Hispanic teachers may help provide insights into more effective means of delivering instruction to students from those ethnic groups.

The challenge for the 1990s is clearly to devise and implement enriching academic programs in mathematics and science for underrepresented ethnic students. It also seems clear that a focus on middle and high schools is not sufficient. Quality programs must be implemented at the primary grades to establish a solid foundation for a greater number of underrepresented ethnic students. The project described here is one attempt at meeting these needs.

The Groups Involved

The activities of the SDMEP involve four groups: children, their parents, their teachers, and college students preparing to be elementary school teachers. Other groups, such as principals, school district administrators and support staff, and representatives of industry, interact with various parts of the project. Figure 3.1 shows the interrelationships between the components of the project.

The Students in the SDMEP

Participation in the SDMEP, by design, was restricted to African-American and Hispanic children. The project started with children in grade 2 and will continue with them through grade 4. From an initial participation of seven classes at three schools (roughly 150 children) during the first year, there was an increase to thirteen classes at six schools (roughly 250 children) during the second year.

The Teachers

The selection of the teachers was very important. Not only would they be devoting many extra hours to the children, but they would also be spending extra time preparing and attending sessions to learn about the sorts of lessons on which the SDMEP would focus. Each principal suggested a small number of his or her faculty who would likely be interested in the after-school work. In particular, an intent of the SDMEP was to involve African-American and Hispanic teachers to develop a coterie of minority teachers with enhanced backgrounds in mathematics.

The SDMEP director then interviewed the candidates individually, hoping to impress on them the potential import of the work for the children and to ascertain their likely ability to meet the time and dedication demands of the SDMEP. This two-step screening has, so far, been quite successful. The teachers in the SDMEP have been remarkably involved and reliable and have been a strong feature of the SDMEP.

College Students

Two groups of college students are involved in the SDMEP. One group, Hispanic and African-American engineering majors, visits classrooms roughly once a week. The second group is composed of college students who are planning to become elementary school teachers. They too are drawn primarily from minority students and thus also serve as role models. Each works with a project teacher on a continuing basis.

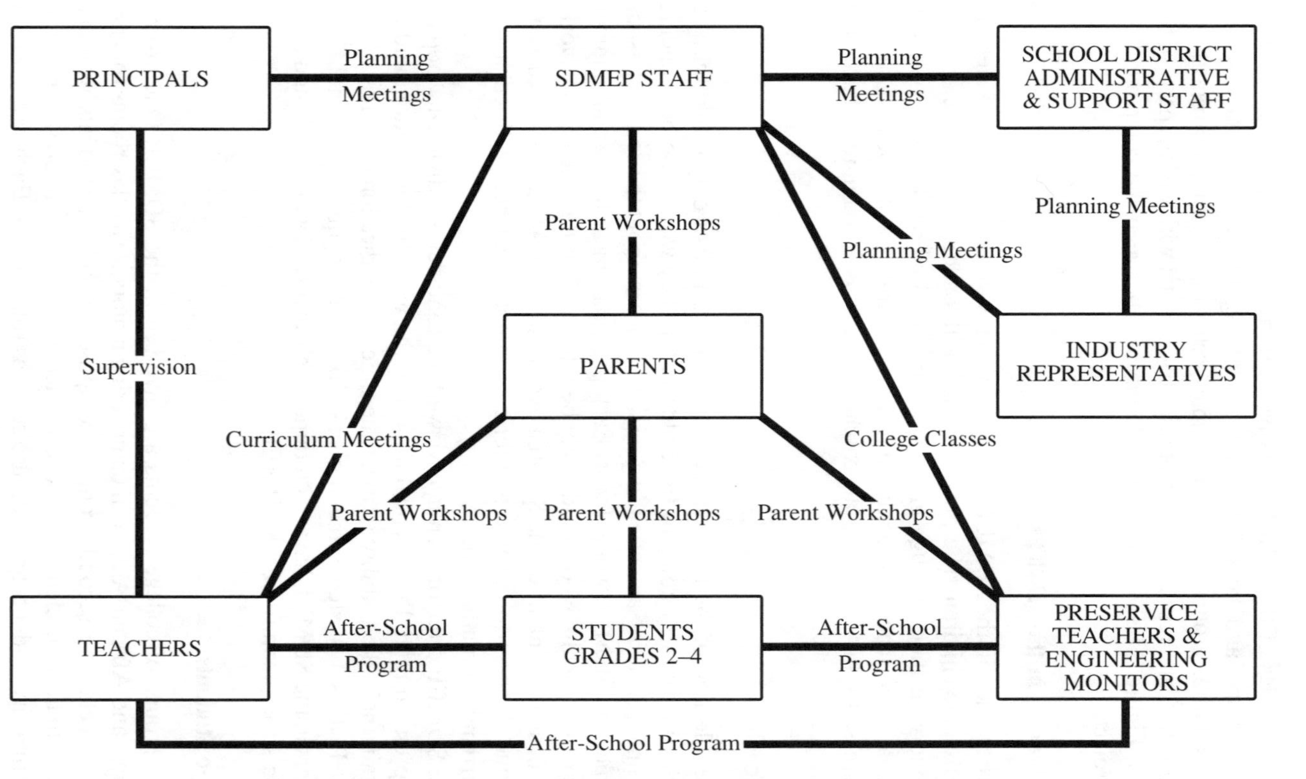

Fig. 3.1. Components of the San Diego Mathematics Enrichment Project

The Parent Council

The importance of parental involvement and support for the success of the SDMEP has been recognized since the program's inception. Emphasis was placed on the creation of a parent-support-and-information network to enhance participation in the achievement of SDMEP goals and objectives. Interest in developing such a network led to the creation of the SDMEP Parent Council.

The council idea was an outgrowth of parents' interest in exchanging information and experiences regarding SDMEP activities in the different participating schools. Many parents felt that an informal organization in which parents could discuss issues related to their children and the impact of the mathematics enrichment experience would facilitate a greater awareness and support for the SDMEP. Another rationale for the council grew out of the recognition that parents should be empowered to develop their own processes of interaction with, and evaluation of, SDMEP activities and objectives.

Other Groups Involved with the Project

Several other groups were important to the growth of the SDMEP. School district administrators and support staff provided direction and resources to support the planning and implementation of the project. Principals provided site-specific coordination, assisting project staff in identifying and communicating with teachers, parents, and students. In addition to their vital role in initiating the project, representatives of industry furnished resources that supported the project. The contributions of all these groups were synergistic in the development of the San Diego Mathematics Enrichment Project.

Activities with Each Group

Activities with Students

At the first meeting of the after-school program, Erica seemed somewhat apprehensive and asked, while eyeing the large cardboard box we had carried in, "When are we going to do the sheets?" Her expectation, at grade 2, was that mathematics lessons meant "doing sheets"! The focus of the project is not on "doing sheets" but rather on conceptual development through hands-on work, exploration, and reasoning. Many of the project lessons do involve paper, but we hope that Erica no longer equates doing mathematics with "doing sheets."

Activities with students include the after-school mathematics enrichment sessions for three hours (spread over two or three days) each week, a summer enrichment program that meets for three weeks for fifteen hours each week, and separate activities at Saturday morning parent workshops (see below). The summer program also includes such field trips as one to the Jet Propulsion Laboratory in Pasadena and a behind-the-scenes look at the San Diego Airport.

As a district-endorsed program, the content of the after-school program supports the district's curriculum; the activities could well be a part of the regular mathematics program if there were time for them. (Some of the project activities are, of course, new to the teachers and give them ideas about ways of teaching similar topics in their classrooms.) Activities are planned with the exciting visions of the *Curriculum and Evaluation Standards* (NCTM 1989) and the California *Framework* (California State Department of Education 1985) in mind. Since the regular curriculum gives a good amount of attention to the development of calculation skills, the after-school work focuses on the central concepts of the mathematics curriculum rather than on computation.

SDMEP has been enriched by generous help from the project directors of several National Science Foundation curriculum projects. Although those projects were not set up to disseminate materials during their development stages, we benefited from early drafts of materials from the Education Development Center (EDC) Journeys in Mathematics Project (available from WINGS, Inc.), the Used Numbers Project (now available from Dale Seymour), the Calculators and Mathematics Project—Los Angeles, the University of Chicago School Mathematics Project, and the Logo-based Geometry Project. Lesson ideas were also drawn from some of the excellent hands-on curricula available for the lower grades, from presentations at state and national mathematics teachers' organizations, and from the wealth of supplementary materials available from commercial publishers. Project teachers usually modify the lessons, of course, or substitute ones of their own creation or from their work in the in-service portion of the project.

A "Typical" After-School Session

The typical day involves a variety of activities, some hands-on, some small-group, and some individual work. The guiding principle is to promote conceptual development with minimal attention to computational skills. Whenever possible, children are asked to verbalize and to justify their thinking. What follows is not a report of a particular afternoon but is an effort to illustrate features of the program.

Most teachers start the session with a snack, not only to meet the children's hunger pangs but also to provide a time buffer for the children to convene from their different classrooms, the restrooms, and the playground. Teachers do not work with children who are in their own classrooms. This practice gives both the teacher and the children a fresh face and a fresh start on the mathematics involved. Each teacher is paired with one college student to allow for more teacher interaction with the students.

Through a signal bell or the teacher's holding up of a hand, which is imitated by the children as a part of the "we're supposed to alert our neighbors, be quiet, and pay attention now" routine, the day's first activity (called Six-Cube City) starts. The teacher holds up a shape made with six cubes (e.g., as in fig. 3.2) and asks the children what they see.

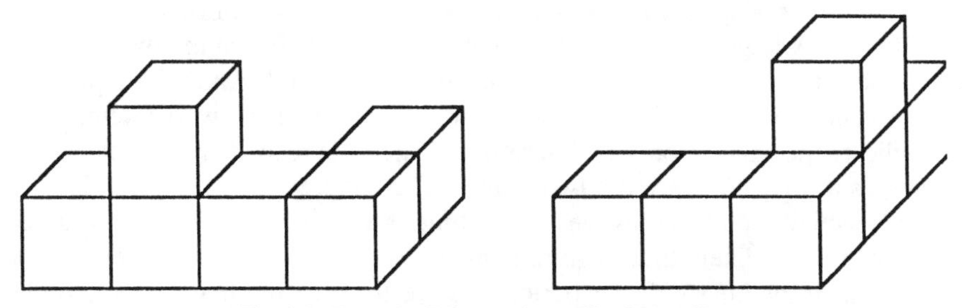

Fig. 3.2. Two "buildings" from Six-Cube City

The teacher uses students' answers to review some key vocabulary from earlier lessons: volume, area, height, length, width, perimeter. Eventually the teacher announces that today the shape will be a "pretend" building with a volume of 6 cubes. The children's job will be to work in assigned pairs to make a Six-Cube City on a strip of butcher paper, according to these "rules," which are written on the board:

On the brown paper, draw a street that is 10 centimeters wide.

Your pair can make about 8 buildings. (The teacher has about 500 cubes, which limits the number of cubes available to each pair.)

Each building must have volume of 6 cubes.

Each building is to be different from the others you make.

Put some buildings on both sides of the street.

Trace around the bottom of each building. Tell what the area of the bottom is.

The teacher and the college student then distribute the materials, and the children start to work. The adults circulate, watching how the children work together, listening to their language use, and asking questions like "How did you know these two buildings were different?" or "Would any of your buildings look the same if you turned them upside down?" The questions are intended to involve the children further in technical communication, to review the principal ideas of the activity, to examine their reasoning, or to present other problems. Whenever possible the problems connect the lesson to other areas of mathematics. For example, the question "What is the total volume of all the buildings that everyone made?" invites the use of multiplication and perhaps addition in this measurement setting. Note that this focus on problem solving, communication, reasoning, and connections is aligned with many of the recommendations of the NCTM *Curriculum and Evaluation Standards* (1989).

The teacher then weighs whether to have the children decorate their cities by drawing driveways, flower beds, the tops of trees, and so on, and perhaps write a story about their city. She decides instead to keep the children in pairs with the following activity, which she has used before with other geometric materials. A large book or stack of books is placed between the two children to shield their work areas from each other. One child makes a shape with cubes and then describes the shape to the other child, who attempts to duplicate the shape from the description. When the second child thinks he or she has it, they compare the two shapes. The children then reverse roles.

An activity that is flexible in its time requirement is often required. Estimating the likely time requirements of work with concrete materials is a skill that sometimes defies experience, since the children may become quite involved (or uninvolved) with some of the activities. Games like "But Who's Adding," "But Who's Counting," and "But Who's Multiplying," all from *Square One TV* (Children's Television Workshop 1989), are often used since they provide an opportunity for students to think about strategies as well as use and enhance their knowledge of place value or basic addition or multiplication facts. It is no longer surprising to encounter a student who does not know the basic facts well but shows quite sophisticated strategies in cleverly blocking the other player or team or in planning ahead for a winning position.

As a closing activity, the children may write in their journals (using whatever language, English or Spanish, with which they are most comfortable) something about the day's work. The teacher and the college student often have to give prompts: "Was there anything today you liked?" or "What did you learn today?" Some children do not like to write, but when they see others writing and are prompted, they usually join in. Occasionally the teacher plans time for the children to read their journal entries to the others.

Activities with Teachers

Two activities are for "teachers only" in the SDMEP: hour-long, biweekly, after-school meetings to discuss upcoming project lessons, and monthly Saturday workshops and a two-week-long summer workshop at which various mathematics topics, such as geometry and problem solving, are discussed in more detail.

The biweekly meetings provide an opportunity for teachers and SDMEP staff to discuss upcoming lessons and nitty-gritty details of the SDMEP, such as the availability of materials and future events. In addition, the Saturday workshops enable teachers to consider their teaching of, and students learning of, various mathematics topics.

Through working with the teachers in the after-school enrichment sessions, we came to realize that the teachers would be more effective and confident if their knowledge of mathematics content and methodology was enhanced. The in-service component was designed to add depth to the teachers' knowledge of elementary school mathematics content by providing meaningful learning activities that allowed them to understand much of what they had learned in a rote fashion. At the same time, methods that allow students to make meaning out of mathematics were introduced to the teachers. The content was organized according to the California *Framework* and NCTM's *Standards* and was presented in conjunction with the type of pedagogy promoted by those documents. Many of the goals set for the in-service component coincide with the "Teaching Standards" in NCTM's (1991) *Professional Standards for Teaching Mathematics.*

In addition to the teachers in the after-school program, other interested teachers were identified through the principal of each school in the target area. The inclusion of a greater number of teachers in the in-service program meant that mathematics learning could be further strengthened for students in the target schools. Other things being equal, preference was given to African-American and Hispanic teachers. Not only were children from these ethnic groups the major groups in the target schools (and could therefore readily adopt such teachers as role models), but the relative scarcity of teachers from such populations is a national problem. As is noted in *Everybody Counts* (National Research Council 1989), "During the next decade, 30 percent of public school children, but only 5 percent of their mathematics teachers, will be minorities. The inescapable fact is that two demographic forces—increasing Black and Hispanic youth in the classrooms, decreasing Black and Hispanic graduates in mathematics—will virtually eliminate classroom role models for those students who most need motivation, incentive, and high-quality teaching of mathematics" (p. 21).

One of the major goals of the in-service program was to provide the opportunity for teachers to gain a more comprehensive view of the structure of elementary mathematics. This would lead to their assuming more responsibility and control over their mathematics teaching and to their becoming better consumers of prepared materials, both district and commercial. Too often elementary school teachers have been led to believe that mathematics instruction is a series of loosely connected lessons that have been adequately prepared and sequenced by textbook authors.

During the first in-service session, teachers were introduced to the NCTM *Standards* and were given a review of the contents of the California *Framework.* These documents were used to paint the "big picture" of the current content and direction of elementary school mathematics and the rationale for educating today's students differently from what was done in the past. The teachers participated in activities that were designed to emphasize the difference between procedural and conceptual learning, and they also took part in some sample activities from the Equals Project (e.g., Downie, Slesnick, and Stenmark [1981]) that stress the issue of equity.

Topics for the subsequent in-service sessions came directly from the *Curriculum and Evaluation Standards* and the *Framework* documents. Topics included number (including developing meaning for whole-number operations, basic facts, and algorithms; introductory work with rational numbers in the form of fractions and decimals; and estimation), problem solving, geometry, measurement, and patterns and functions. Each in-service session started with more specific references to the NCTM *Standards* and the *California Mathematics Framework* and how the session's content fit into the overall picture of the elementary school curriculum. Relationships between mathematics topics were constantly stressed. The idea that mathematics learning is analogous to a weaving or structure of the learner's own creation was presented to the teachers as a way to promote their thinking of the interrelatedness of mathematics content and mathematics learning experiences. Theoretical bases were discussed, and activities were presented as examples of how the theory could be put into practice. Teachers were encouraged to modify, expand, or reject activities as they considered the needs of their particular students.

Research has expanded our knowledge of how students learn mathematics. Although teachers need access to this research, they often do not have the time to read and interpret research journals on a regular basis. Research findings were presented to the teachers when appropriate and as they applied to each topic of discussion. One might say that the in-service approach was holistic in nature and gave the teachers the opportunity to construct their own knowledge and belief system about teaching mathematics.

A "Typical" In-service Session

Although the content of the sessions varied, the structure remained fairly stable. The following description gives a sense of how the monthly Saturday workshops were organized and presented.

The teachers arrive with their arms full of munchies and containers of coffee. They immediately start to catch up with colleagues on personal and professional information and to share with the in-service leader individual anecdotes from their classrooms. The anecdotes center on the implementation of ideas gained from the in-service program. For example: "I did those multiplication patterns on the ten-point circle, and the children loved the activity. We decided to call them Magic Circles and we hung them up for open house." The teachers, many of whom just spent three days from the previous week doing the after-school enrichment program as well as teaching full time, are amazingly eager to learn more about mathematics.

Two content topics are planned for the day, geometry and measurement. The topics are closely related, and it is the in-service leader's goal to show that relationship. The session begins with a broad view of how geometry fits into the elementary school curriculum. References are made to the *Standards* and the *Framework* and major points from each document are summarized. In the introduction the instructor stresses that the study of geometry in the elementary school begins in kindergarten and continues throughout the elementary years and furnishes the opportunity for rich language experiences, aesthetic visual expression, and connections to the child's real-world experiences.

Throughout the morning, learning activities are introduced that promote language development through speaking, reading, and writing, hands-on experiences, recognition and generation types of tasks, and connections to the children's own background experiences. The van Hiele levels (e.g., Crowley [1987]) of geometric thought serve as a guide for the in-service leader, but they are not introduced to the teachers at this time.

The teachers are presented with a sequence for teaching geometry that starts with solid shapes, then moves to plane shapes, and later includes the introduction of more abstract geometric ideas such as points and lines. The teachers are made aware of the fact that the sequence is a theoretical one, not one that has been thoroughly researched for effectiveness. The sequence is intended to build on the childrens' extensive experiences with three-dimensional objects in the form of toys and household objects. Such familiar objects as shoe boxes, balls, and soft-drink cans are related to models and pictures of geometric solids. Teachers nod in agreement as the point is made that textbook pictures of three-dimensional objects are often too abstract for children unless deliberate connections are made between pictures and objects. Language furnishes students with a way of connecting these experiences. Also introduced to the teachers as additional resources are examples of children's trade books that illustrate geometric shapes, contain text about geometry, and provide geometry activities.

Students need a combination of generation and recognition kinds of tasks to form solid concepts. Recognizing different examples of cubes, for example, is an important type of task. Generating examples of cubes by constructing them from two-dimensional patterns adds to children's understanding of the concept of cube and readies them for relating solid shapes to plane shapes. The teachers' comments as they construct examples of geometric figures reveal their own growth in understanding as they feel cardboard edges, vertices, and faces and realize that they are feeling lines, points, and planes.

Teachers are reminded to look around the classroom in order to make connections between the mathematics they are teaching and the real world the children live in every day. The room itself is used as an example. The teachers classify it as a solid, further classify it as a rectangular prism, and then go about examining the edges, vertices, and faces of this giant prism from the inside out. Labels are made and stuck to the walls. Geometry surrounds the group.

Activities for the rest of the morning continue in the same vein as the teachers revisit plane geometry and the study of more abstract figures, such as points and lines. Constant attention to the connections between the topics and their interrelationships are stressed. A shoe box becomes a model for a solid shape that is a polyhedron, is also a rectangular prism, and has faces, edges, and vertices. The box can be dipped in tempera paint and prints made of the faces. The edges print as lines, and the vertices print as points. The teachers' faces reveal their wonder and excitement at the new and exciting ways of making mathematics come alive for their students.

Measurement is the afternoon topic. The in-service leader points out from the beginning that the geometry and number strands (among others) intersect in the measurement strand. Students better understand area measurement if they understand the idea of a plane surface, and they are better able to measure to the nearest fourth of an inch if they understand and have experience with locating and recognizing fractions on a number line. Often these topics are not sequenced this way in textbooks, and the teachers immediately start listing examples with a better understanding of why their students were not successful with particular lessons.

Activities and materials for the in-service workshop were gathered from a variety of commercial sources. Resource books for such manipulative materials as pattern blocks, geoboards, and tangrams were used. Activity books for students that contain nets (patterns) of solid figures to construct provided generative activities. Marilyn Burns's (1988) videotapes furnishes examples of geometric problem solving and higher-order thinking through activities and questioning techniques for teachers.

Activities with College Students

The SDMEP also involves two groups of undergraduate college students: engineering majors and prospective elementary school teachers. The engineering majors visit classrooms roughly once a week, tutoring individual students in mathematics and providing role models to the students in the SDMEP.

The prospective elementary school teachers are assigned to a single classroom, where they work with the teachers in planning and delivering the after-school activities. These college students assist in the instruction, often leading small-group activities. They also attend weekly sessions at the university (for which they receive course credit) at which the upcoming lessons in the after-school program are discussed and that also provide a forum for them to discuss any questions or concerns they have regarding their participation in the SDMEP.

Activities with Parents: The Parent Council

Successful educational programs focused on enriching the learning experiences of children require the support and participation of parents. The SDMEP experience demonstrates that parents who feel empowered through involvement in the operation and evaluation of such programs can make substantive contributions to their success. The institutionalization of processes of parent involvement suggested by the Parent Council can ultimately lead to newly organized grass-roots advocacy for the continuation of educational innovations at all educational levels.

The main objective of the parents participating in the SDMEP is to enhance the educational enrichment of their children. The Parent Council has become the focal point for assessing parent interests and concerns about the processes and progress of such enrichment. Council activities have taken several forms. Council members have positioned themselves strategically among school-site administrators, clerical staff, and parent systems. They have developed telephone "trees" for disseminating information quickly and have demonstrated their success in many of the activities. The Parent Council also has instituted a quarterly newsletter that is distributed to all parents, teachers, and project staff.

The council has spearheaded parental support for SDMEP teachers, college students, and curricular innovations. Its members have done so by visiting after-school sessions, participating in parent workshops, and recommending to other parents that they do so as well. Council members have played the important role of liaisons and contact points between SDMEP schools and their broader parent constituencies regarding special school events.

Council members also have planned monthly parents' meetings at each SDMEP school. These meetings have served as forums for informing parents about various activities, ranging from those in the after-school sessions to the project's field trips. The meetings have also focused on parents' questions and concerns regarding their children and the SDMEP. Parental input has resulted in changes in the projectfor example, in the format and content of the parent workshops. Parents expressed genuine surprise that their requests and suggestions had been implemented. For many, it was the first real experience with parental empowerment.

SDMEP parents have recently articulated concerns regarding the prospect that the project will end within approximately one year. Expressions of this concern through the council have begun to take on the form of an organizing campaign to continue and expand the project's activities to other schools and grade levels. Council representatives are developing strategies for dealing with the SDMEP's future.

One member's comment serves as an adequate summary of parental concern: "This is the kind of program our children and community need, and the school district and board need to know that we want it continued.... Whatever it takes, we want it continued."

Parent Workshops

The parent workshops are held two to four times a year at a community college centrally located to the elementary schools involved in the SDMEP. The parents of all the students in the SDMEP are invited to attend. These workshops have several purposes: to provide an opportunity for parents to participate in activities similar to those that their children do in the after-school mathematics sessions, to familiarize them with current directions in the teaching of mathematics (e.g., that mathematics is more than just computation), to learn more about how they can help their children do better in mathematics, and to improve communication among all persons involved in the SDMEP. Incidental learning often results in the strengthening or updating of parents' knowledge and understanding of mathematics.

A "Typical" Parent Workshop

Parent workshops are held on Saturday mornings and are attended by parents and their children as well as the teachers and college students involved in the SDMEP. Activities begin with a Continental breakfast while participants are signing in. Next, all participants convene in an auditorium, where SDMEP staff welcome the participants, present an overview of the morning's activities, and answer any questions. Following this overview, the groups separate, with students, teachers, and college students working together while the parents meet in another room for separate activities. Later in the morning, these groups will reconvene; parents and children will participate in mathematics activities together. The morning ends with a wrap-up, with parents completing an evaluation of the workshop.

The goals of the session with parents are to broaden their view of mathematics and mathematics learning and to furnish them with examples of several activities similar to those done in the after-school mathematics sessions that they could do with their children at home. To broaden parents' view of mathematics and mathematics learning, this session began by viewing a videotape entitled *Math Matters: Kids Are Counting on You,* prepared by the National Parent-Teacher Association (1989). After watching the tape, parents discussed their opinions regarding common misconceptions related to mathematics, such as the following: "Math is just arithmetic," "Math isn't important," "Math is really boring," "Math is too hard for everyone to learn," "It doesn't matter how hard you work; if you're not good at math, you'll never get any better," "I was always bad at math, so my kids won't be good at it, either." Many parents seemed to enjoy the opportunity to share their feelings and experiences as learners of mathematics and also to discuss how their attitudes have changed.

Continuing to help parents recognize that mathematics is not just numbers, we briefly discuss with parents the California *Framework* and the NCTM *Standards.* At other parent workshops, we have discussed ways to help their children do better in mathematics, including setting up a designated time and area at home for them to do their homework.

The next part of the parents-only session is devoted to parents' participation in activities similar to those done in the after-school sessions. We discuss the potential benefits of each activity and how they fit into the "big picture" of what mathematics is. This session's activities include work with tangrams, logic activities with squares of colored paper, and number and logic activities from *Square One TV* (Children's Television Workshop 1989) entitled "But Who's Adding" and "But Who's Multiplying," which are two of the children's favorite games. Each parent receives copies of the activities and the materials needed to do the activities with their children at home, including colored paper squares, sets of tangrams made of tagboard, and the playing boards for the games.

Effects of the Project

Information describing the effects of the SDMEP was obtained in several ways: through an analysis of test data; individual interviews of thirty students, questionnaires to parents, teachers, and college students, and observations by SDMEP staff.

Effects of the SDMEP on Students

A preliminary analysis of students' scores on the Comprehensive Test of Basic Skills (CTBS) shows a higher percentage of SDMEP students scoring at or above the publisher's median at all the schools that have been involved in the SDMEP for one school year. Further information on CTBS performance is difficult to obtain because of high student mobility and confidentiality of test scores.

Students' comments about their own learning. Several of the students interviewed commented that they were doing better in mathematics because of their participation in the SDMEP:

- "I didn't know how to do my times tables then (before the SDMEP). I learned them in after-school math."
- "When I was in the first grade or in the second grade I did bad in take-away's, and then when I came here I did a little better. They show us how to do it, and if you don't know the answer then they help you with it."

Students' comments about their attitudes toward mathematics. Several of the students remarked that they now liked mathematics:

- "I like it (math) and I keep on working on it because it's fun."
- "You can be playing a tens and ones game, and you could still be learning and you won't even know you're learning 'cause you're having so much fun."

Other students' comments about the SDMEP. Students made other comments regarding their opinions of the SDMEP:

- "After-school math is real nice and you can really learn there. I feel proud of myself."
- "I think that after-school math is the best thing that happens in my life."

Teachers' comments on students' learning. The SDMEP teachers were asked to describe the effects the SDMEP has had on the participating students. Teachers reported that the students involved in the SDMEP have made great progress as a result of their participation. They have progressed in many areas, including mathematics achievement and understanding:

- "I think they have a clearer understanding of what is really happening when they perform certain operations."
- "Concept mastery in many instances seems to take less time."

Teachers also reported changes in students' attitudes and confidence about their ability to do mathematics:

- "Students are eager to solve problems using manipulatives. They appear to have more confidence in their ability to solve problems. They think math is fun."
- "The students are more interested in taking a chance. I see more students' not being afraid to problem-solve."

Teachers reported that many of their students have more perseverance when solving mathematics problems and that students now are associating mathematics with many other things in everyday life.

- "Students have been very involved in thinking skills as a result of this project. Students tend to work on problems and strategies longer as a result of this program."

Other teachers commented on the connection between the growth in their own understanding of mathematics and the improvement of their students' learning:

- "They (the students) have grasped many new concepts and skills through the use of the manipulatives and instruction I have received."

The teachers reported both cognitive and affective gains in participating students.

Parents' comments regarding changes in their children. Parents were asked if they have noticed any changes in five selected areas in their children enrolled in the program. Most of the parents reported that their children like mathematics more, have better study habits and improved writing skills, and are more self-confident and feel better about themselves. Their responses are as follows:

	Percent responding			
	Yes	Not sure	No	No response
• Like math more	82%	8%	0%	10%
• Have better study habits	80	6	2	12
• Are more self-confident	78	8	2	12
• Have better writing skills	73	12	2	12
• Feel better about themselves	69	12	2	16

(Note: $N = 49$)

Other changes in children. Many parents commented on other changes that they noticed in their children. These changes include better mathematics understanding and achievement, better attitudes about mathematics, more confidence in their ability to do mathematics, better study habits, and a better attitude toward learning. The following are selected comments:

- "My son talks a lot about going to college. He talks about being a doctor and an inventor."
- "My child has demonstrated more confidence in his abilities to do math. He looks forward to attending class."
- "My child had a fear in mathematics in learning times tables, division, etc. However, I have noticed a change in my child: she feels confident, assured, she knows her times tables, and it is the result of this program."

Effects of the SDMEP on Teachers

"I have tremendously changed my style of teaching math, and I feel my students are benefiting from it." That comment was made by one of the SDMEP teachers when asked to describe the effects the SDMEP has had on them as teachers. Teachers reported that they have changed *what* mathematics they teach, *how* they teach it, *how* they evaluate students' learning, and *how they feel* about teaching mathematics because of their participation in the SDMEP. As part of the SDMEP, teachers not only broadened their content knowledge but were encouraged to think about and discuss their own learning processes.

Changes in what mathematics they teach. Teachers are teaching more than just computation in their mathematics classes.

- "I try to touch on several strands of math each week to eliminate the idea that math is just computation."

Changes in how they teach mathematics. Teachers reported that they use manipulatives more often in teaching mathematics, that they are better able to diagnose and meet students' needs, and that they are aware of more strategies to use in teaching mathematics:

- "I've gained greater insight into strategies I can use to broaden their understanding.
- "I am more willing to use manipulatives in my regular math class."
- "I am much more aware of the importance of manipulatives and making the students think."

Changes in how they feel about teaching mathematics. Teachers are having fun teaching mathematics:

- "I have learned that good math teaching is effective modeling of all concepts and taking the kids where they are, helping them to become active learners, critical thinkers, and number explorers, and to develop a can-do attitude about math."
- "You definitely need to use manipulatives with all age groups."
- "I am a completely new teacher when it comes to math. I now have a totally new idea of what good teaching is all about."

View of what mathematics is. Some teachers commented that their view of what mathematics is and their attitudes toward mathematics have been changed because of their involvement in the SDMEP:

- "Math is a process for me now instead of filling in worksheets, etc."
- "I enjoy math much more."
- "I am truly understanding math along with my students and college student. We are learning that we can have success in math."

The most effective part of the project. SDMEP teachers listed the components of the SDMEP that they felt were of critical importance to their involvement. These factors include small groups, readily available lesson plans and materials, supporting teachers with a salary stipend, teacher enhancement component, parent involvement, parent-teacher-child group meetings, and close association with the university.

Effects of the SDMEP on College Students

The college students involved in the SDMEP felt that their involvement had benefited them in many ways. The students felt that because of the project they had a better idea of what it's like to be a teacher and are finding more relevance in their university coursework. Other students commented that the experience of working in the SDMEP helped them to become more aware of what it was like to teach learners of different abilities and also increased their confidence in their understanding of mathematics, which would help them be better teachers.

Effects of the SDMEP on Parents

Many parents are extremely supportive of the SDMEP's activities. Many parents reported that they have changed their attitudes about mathematics and their school-related interaction with their children. Several parents have suggested expanding the program's activities with children and with parents.

Parents were asked if they are doing anything different as a result of the program. Twenty-five (51%) of the forty-nine parents responded affirmatively. Most of these parents are spending more time working with their children at home. The following are selected comments:

- "I have tried to create a quiet time each day for them."
- "I continue to express to my child the importance of math, to take advantage of this opportunity to enhance his skills."
- "Since I've been introduced to the program, I have created a math environment in my home and put emphasis on the math in other activities we experience together."

Other changes in parents. Parents also reported changes in their attitudes about mathematics and their awareness of the importance of mathematics and ways they can help their children learn mathematics. Some parents are taking classes to learn more mathematics. The following are selected comments:

- "I used to hate it and avoid it. Now I'm learning it can be fun."
- "I'm more aware that you must start earlier with children with math and continue to work with them daily."
- "I like it very much. Im even learning myself."
- "I now have enrolled in College Algebra after eight years of no school."
- "I feel if I learn more, I can help him when necessary."

Favorite part of the parent workshops. Parents appreciated the mathematics activities to be done at home with their children as well as the opportunity to learn more about mathematics.

- "Parents learning more about math. You could really see the interest and excitement."
- "They teach us how to work in math with kids at home."
- "Activities working with different mathematical skills that we can do as a family."
- "Learning new games because we play them five days a week and I needed more. Thanks."
- "Learning more ways of demonstrating math to my daughter so that she will find each enjoyable."

Students also reported that the SDMEP has had an effect on the things they do at home with mathematics.

- "More time. I do more of it, every chance I get. We play 'But Who's Multiplying?' and 'But Who's Adding?' and 'Guess my Number.' My sister and I play, only sometimes with my parents."
- "I play games at home like 'Race to a Hundred.' I use flats and longs and units. Some Saturdays we go to the workshops, and they let us take home some (paper base-ten blocks)."

Suggestions for Replication

Planning Such a Program?

Some points to consider when planning an after-school program like this one follow.

- *Select the teachers carefully.* It is obvious that the teachers are the key to the success of such a program. The program makes surprising demands on their time and hence requires considerable dedication.
- *Furnish adequate teacher in-service support.* Even teachers of good will, like the ones in our project, may need some updating in content or advice on how to implement some of the recommendations of the NCTM *Standards.* Our teachers fortunately also had the time and energy to participate in the in-service portion of the project, which enhanced their confidence noticeably.
- *Support the teachers and students in as many ways as possible.* Methods of support can include stipends for teachers, prepared lessons, available manipulative materials, resource books, supplies, field trips, snacks for students, and an end-of-year recognition ceremony. It is essential to obtain the support of the mathematics coordinator and the school administrators, since they often provide resources.
- *Build in teacher input and interaction.* Teachers need time to furnish input into the lesson development and revision as well as to make suggestions regarding organizational details of the project. Teachers also need time to share reactions and experiences with other teachers—for example, discussing ideas that worked or did not work, what variation paid off, or what supplementary or alternative activity seemed particularly promising.
- *Involve parents and teachers in project planning.* Their input is critically important in producing a high-quality project and in maximizing the project's impact.
- *Give suggestions to the parents about things to do at home.* Many parents *want* to help their children at home but do not know what sorts of things might be appropriate or supportive. Sometimes the parents will not feel secure enough themselves to work with their children, especially when some of the current topics, like probability, the metric system, and geometry, are ones to which they have had little exposure. Our parent workshops have helped along these lines.

- *Coordinate all components of the project.* Good communication among all persons involved is vital to the success of projects. A master calendar, preferably for the entire school year but at least for one semester, aids in this coordination. Telephone calls by teachers to parents, as well as initial door-to-door, in-person contact, are also helpful. We are currently instituting an SDMEP newsletter containing schedule information, information on topics covered in after-school sessions, and additional ideas for at-home activities for parents and children to do together to extend and reinforce SDMEP activities. Often it may be possible to involve a district's staff development resources in the planning and support of such a project.
- *Involve at least a "critical mass" of teachers from each school involved.* Project staff should hold a meeting at each school explaining to all teachers the need for the project, its purpose, activities, and the benefits to them and the students.
- *Involve college students in the project.* Form teams of one or two college students with each teacher at the beginning of the project; keep these teams intact for at least one school year.
- *Use all resources available.* For example, some parents were eager to serve as aides in the after-school sessions and districts often have mathematics resource teachers available who can make valuable contributions to project activities.

"I'd like to thank everyone involved in the program for giving us all something to be proud of—our children and ourselves." One goal of the San Diego Mathematics Enrichment Project is to help all students, teachers, and parents achieve what one parent describes here—to be proud of themselves in relation to mathematics. It seems that at least to some extent the SDMEP has begun to achieve this goal.

References

Berkeley (California) Unified District. *1983 CTBS Test Scores.* Berkeley, Calif.: The District, 1983.

Burns, Marilyn. *Mathematics: With Manipulatives.* Videotape. New Rochelle, N.Y.: Cuisenaire Company of America, 1988.

California State Department of Education. *Mathematics Curriculum Framework for California Public Schools, Grades Kindergarten through 12.* Sacramento, Calif.: The Department, 1985.

Childrens Television Workshop. *Square One Television Game Shows Teacher's Guide.* New York: The Workshop, 1989.

Crowley, Mary L. "The van Hiele Model of the Development of Geometric Thought." In *Learning and Teaching Geometry, K–12,* 1987 Yearbook of the National Council of Teachers of Mathematics, edited by Mary M. Lindquist, pp. 116. Reston, Va.: The Council, 1987.

Downie, Diane, Twila Slesnick, and Jean K. Stenmark. *Math for Girls and Other Problem Solvers.* Berkeley, Calif.: Math/Science Network, Lawrence Hall of Science, University of California, 1981.

Matthews, Westina. "Influences on the Learning and Participation of Minorities in Mathematics." *Journal for Research in Mathematics Education* 15 (March 1984): 84–95.

National Council of Teachers of Mathematics. *Curriculum and Evaluation Standards for School Mathematics. Reston, Va.: The Council, 1989.*

———. *Professional Standards for Teaching Mathematics.* Reston, Va.: The Council, 1991.

National Parent-Teacher Association. *Math Matters: Kids Are Counting on You.* Videotape. Chicago, Ill.: The Association, 1989.

National Research Council. *Everybody Counts*. Washington, D.C.: National Academy Press, 1989.

Quality Education for Minorities Project. *Education That Works: An Action Plan for the Education of Minorities.* Cambridge, Mass.: Massachusetts Institute of Technology, 1990.

Rothman, Robert. "Blacks, Hispanics Lag in Math by 3rd Grade." *Education Week*, 3 August 1988, p. 7.

4

Multicultural Mathematics: One Road to the Goal of Mathematics for All

Claudia Zaslavsky

As I look back on my teaching experiences in Greenburgh Central Seven, a school district just north of New York City that was known nationwide for its decision in 1951 to integrate its schools by busing, I am struck by the similarity between the problems we encountered in the sixties and seventies and those of today. The solutions we worked out may be of value to teachers trying to reach all students with mathematics. In particular, this chapter will deal with the introduction of elements of a multicultural mathematics curriculum in the middle and secondary grades.

The National Council of Teachers of Mathematics recognizes the relevance of culture: "Students' cultural background should be integrated into the learning experience" (NCTM 1989, p. 68).

In a chapter contributed to the book *Empowerment through Multicultural Education,* James Banks, one of the foremost authorities on this subject, speaks of the body of knowledge that we expect all our citizens to master (Banks 1991, p.126):

> The knowledge institutionalized within our schools . . . . should reflect the interests, experiences and goals of all of the nation's citizens and should empower all people to participate effectively in a democratic society.

Multicultural education, therefore, should include the contributions of all peoples as well as concern for their problems and difficulties. Further, it should embody the consideration of the factors in our society that prevent the effective participation of all citizens in a democratic society.

The decade of the sixties was a period of ferment and change. One effect on our school district was the students' demand for courses in African history and Swahili, courses that were relevant to their culture. The faculty, too, was offered an optional course in African history, a subject that generally received inadequate, if any, treatment in the college curriculum. For my term paper in the course I decided to research the development of mathematics in Africa south of the Sahara, but I was amazed to discover that there was no heading in the library catalogs for African mathematics or for any of its branches.

Mathematical practices and concepts arose out of the real needs and interests of people in all societies, in all parts of the world, in all eras of time. All peoples have invented mathematical ideas to deal with such activities as counting, measuring, locating, designing, and, yes, playing, with corresponding vocabulary and symbols to communicate their ideas to others (Bishop 1988, Gilmer 1990). I was convinced that African peoples were no exception and that such information must be available. My research eventually led to the publication of a book on the subject (Zaslavsky 1979a). How I used this knowledge in the classroom will be discussed in the latter part of the chapter.

A New Secondary School Mathematics Curriculum

Although our secondary school had few dropouts—partly because of the availability of an adequate number of guidance counselors, who stayed with the same students throughout their high school careers—we found that many students were doing poorly in the traditional academic mathematics courses and were dropping the subject as soon as they had fulfilled the minimum requirements. These students were predominantly of working class background, both black and white. This was not an acceptable development in a district that prided itself on offering the best possible education to *all* its students.

Like the situation today, the decade of the sixties was a period of great concern about the status of mathematics education in this country. The Soviet conquest of space with the launching of Sputnik in 1957 sent shock waves through our government and business establishment, resulting in increased funding to upgrade mathematics and science education. Our district, ever on the alert for such opportunities, obtained funds to write a new curriculum to attract these potential math-dropouts. We were fortunate, too, in that the community supported us as we increased the number of courses offered by the mathematics department, a number exceeding that available in some larger and wealthier districts nearby.

Our mathematics department developed a complete curriculum for grades 9–12, designed to be user-friendly and combining hands-on activities with many of the topics usually covered in grades 9–11. The twelfth-grade unit was based on elementary probability and statistics. The entire curriculum included more computation than is usually found in academic mathematics courses at this level, presented in a manner that would enable students to understand the underlying concepts and some aspects of number theory. As much as possible, applications were derived from local issues and the real world.

I will describe briefly the most relevant topics in the twelfth-year course, then discuss more fully one of the most successful projects, involving census data for the local area. The topics listed below were not necessarily included every time the course was given. Flexibility was a great advantage in designing our own courses. We could add what seemed relevant at the time and omit the topics that did not work well. The topics included the following:

- Probability theory, involving experiments with coins, dice, thumb tacks, and playing cards and a discussion of mortality tables, lotteries, and heredity
- Statistical distributions based on students' data—heights, birthdays, number of children in the family
- Descriptive statistics (tables and graphs) relating to such topics as the integration of whites and "nonwhites" in housing, the amount of nicotine in different brands of cigarettes, the lifetime dollar worth of an education, the comparison of education and earnings of males and females, and the comparison of blacks and whites with regard to education, income, infant mortality, life expectancy, and other factors
- Computation, significance, and applications of the mean, median, and mode; histograms and the normal distribution curve
- Computation of grade point averages, the consumer price index, and the cost-of-living increase incorporated into the contract of the United Automobile Workers union

Certain topics evoked a great deal of interest. This was the period of the Vietnam War, and young men were very concerned about their futures. A Department of Defense publication showed that in 1967 Negro [*sic*] men composed 9 percent of the armed forces but 11 percent of all men assigned to Vietnam and 15 percent of the deaths in Vietnam. The students discussed the reasons for the discrepancy, illustrating their arguments with stories about friends and relatives.

Another interesting set of statistics involved the difference between the cost of a shopping basket of food in the low income and the middle class neighborhoods of New York—poor people paid more for their food than higher income people! Langston Hughes's lines about Harlem were appropriate:

Uptown on Lenox Avenue
Where a nickel costs a dime.
(from "The Panther and the Leash")

Analysis of Data about the Community

The most empowering and, therefore, the most successful lessons were related to the students' own community. Using the Department of Commerce publication *Standard Metropolitan Statistical Areas Outside of New York City, 1960 Census,* I drew a map showing the five Standard Metropolitan Statistical Areas (SMSAs) that composed the school district. I leafed through the book, selecting the features that I thought would be of greatest interest to the students, and copied the data for each SMSA. Among the categories I used were these:

Total population, classified as white, Negro, other

Education and median family income for each category

Male and female employment and unemployment rates for whites and nonwhites

Housing—owned or rented, condition, value, number of persons per room—for whites and nonwhites

Automobiles owned

Every student received a copy of the map and the data. After some discussion, the students paired off, each couple taking responsibility for comparing the five SMSAs from the point of view of two or three characteristics. We had some general discussion about making tables and graphs to display the data, and then they got to work. Although they worked in pairs, each student was responsible for producing the tables and graphs.

I had assumed that this was a simple assignment, but I was wrong. For some, planning the scale for the graph was a formidable task. Their first attempts revealed how little they understood. The scale did not fit on the paper. They did not know how to label the intervals after they had marked them on the vertical axis; if the data gave median incomes of $2320, $4350, $5489, $4190, and $3085 for the five SMSAs, those were the numbers placed next to the marks on the vertical line. Several students were successful only after three or four attempts.

Then the class came together to compare and analyze their productions. With the boundaries of each SMSA clearly marked on the map, each student knew exactly where he or she fit into the total picture. It was evident that the predominantly white, middle class population had far higher incomes, better housing, more education, and all the other attributes of a higher standard of living than the more poorly endowed black and working class people. Of course, the students had been aware of these differences before, but here were the numbers, and numbers don't lie. Here were the graphs, and a picture is worth a thousand words. These students were carrying out one of the goals put forth in the *Standards:* they were learning to communicate mathematically (NCTM 1989, p. 9)!

Most of the students were taking sociology as their senior year social studies course. The social studies department considered me an honorary member of their department, and they welcomed the input from my classes as a basis for further discussion in the sociology course. We were truly making connections among the content areas, as recommended by the *Standards,* and the students were learning to "appreciate the role of mathematics in the development of our contemporary society and [to] explore relationships among mathematics and the disciplines it serves" (NCTM 1989, p. 5).

Nor did parents object to our tackling such a controversial subject as societal inequities. In a district that had opted for school desegregation, we could generally count on the support of the majority of the community in dealing with such topics. Many families, both black and white, had chosen to live in this community precisely because it was less segregated than most of the surrounding area.

Is it wise to deal with controversial social topics in the mathematics classroom? Some British educators have been doing it for many years. In an article in *Mathematics Teaching,* Anthony Cotton (1990, p.24) calls for

> an anti-racist approach which addresses the needs of all pupils in helping them to question values, have a truer understanding of the worldwide sources and applications of mathematics, and gain true equality of access to the curriculum.

Such applications of mathematics touch students' lives and empower them to participate effectively in our society. Through democratic debate students may arrive at an understanding of points of view that are different from their own, thus learning to appreciate the issues that affect the lives of our diverse population. As James Banks remarks (1991, p.127):

> The knowledge that students acquire.... must describe events, concepts and situations from the perspectives of the diverse cultural and racial groups within a society.

Both the faculty and the students considered these courses successful in overcoming math anxiety and math avoidance. We tried to eliminate the need for remedial courses at the high school and college levels by incorporating a review and by reteaching, if necessary, concepts from arithmetic in a context that enhanced the students' self-esteem, rather than labeling them "dumb" and increasing their fear of mathematics. Furthermore, these topics in arithmetic took on more meaning than if they had been studied and drilled out of context (Brown, Collins, and Duguid1989; Janvier 1990).

Many of these potential dropouts from mathematics continued to take mathematics courses throughout their high school years. We were particularly successful with black and working class young people, those groups that had been "underserved and underrepresented" in mathematics (Research Advisory Committee 1989; Reyes and Stanic 1988). Young women were not a problem—females and males in our district were equally represented in advanced mathematics classes as far as the precalculus level—but we did make "conscious efforts ... to encourage all students, especially ... minorities, to pursue mathematics" (NCTM 1989, p.68).

If I were teaching a similar course today, I would recommend books like these by Schwartz (1989) and Frankenstein (1989) as texts for the twelfth-year program. Somewhat less sophisticated are works by Zaslavsky (1988), based on an analysis of the United States military budget, and Zaslavsky (1987), which includes several activities dealing with societal issues, such as infant mortality rates, population changes, and federal expenditures, as well as lessons that incorporate the practices of various cultures. Hudson (1987, 1989) has developed and refined a computer data base containing twenty categories of information about each of 127 countries, with a series of problems to accompany it.

The Mathematics of Various Cultures

In the early seventies I began to explore the possibilities of introducing into the curriculum the study of mathematics as a cultural product, the kind of mathematics developed by various societies to satisfy their own needs. My focus at the time was on societies in Africa south of the Sahara (Zaslavsky 1979a); later I broadened my scope to include some of the underrepresented societies throughout the world. In recent years the mathematical and scientific developments of non-European cultures have become the focus of study for a number of scholars (see Joseph [1991], for example). Not to include such contributions is to imply that these people had no mathematics or science. The same applies to women of all races and ethnic backgrounds. To quote James Banks (1991, p.127):

> The Western-centric and male-centric canon that dominates the school and university curriculum often marginalizes the experiences of people of color, Third World nations and cultures, and the perspectives and histories of women.

I took advantage of a sabbatical year to develop and test materials for several grade levels, based on the infusion of African mathematical ideas into the curriculum (Zaslavsky 1979b, 1981, 1989, 1991). First, I will describe my experiences with such curricular ideas in the secondary grades.

In our district the ninth-grade social studies curriculum included the study of Africa. The two teachers who team-taught this large, heterogeneously grouped social studies class were perfectly happy to allow me to take over their class for a week. On the first day I showed slides illustrating mathematical practices that several African peoples had developed in the course of their everyday activities. The following day I introduced a unit about networks, on which we spent the remainder of the week.

A network consists of a set of points (*vertices*) and the paths (*edges*) connecting them. Figure 4.1 illustrates my favorite network, a figure that was drawn in the sand by elders of the Chokwe people of Angola and Zaire to illustrate their myth about the beginning of the world (Zaslavsky 1979a, 1981). Ascher (1988) and Gerdes (1988) have developed this mathematical topic extensively.

These African networks are examples of mathematical graphs. Graph theory is a growing field of mathematics and has important applications in modern technology relating to communications, highways, and so on. The task I set the students was to trace the given networks and to determine which designs were traceable—that is, could be drawn in one sweep of the pencil, without either lifting the pencil or going over the same line segment more than once. Then they were challenged to determine the necessary conditions for traceability. The project was open-ended; some students were able only to trace the given networks, others drew traceable networks of their own invention, and still others tried to respond to the theoretical challenge. As they indicated on their evaluation sheets, it was a positive experience for all, including the teachers, who later told me that several of these students had not participated in class since the beginning of the school year. They had rejected the papers I distributed the first day, but when they saw their classmates engaged in hands-on work and animated discussions, they, too, wanted to participate. As stated in the *Standards,* "learning occurs through active . . . involvement with mathematics" (NCTM 1989, p. 9). For further discussion of this experiment at both the sixth- and the ninth-grade levels and students' evaluations of the project, see Zaslavsky (1991).

Fig. 4.1. A Chokwe network drawn in sand

Some time in the seventies we adopted Jacobs (1970) as the textbook for the eleventh-year course in the sequence described at the beginning of this article. I soon realized that the first chapter in Jacobs dealt with the same mathematical topic as the networks drawn in the sand by Bakuba (Zaire) children in imitation of their parents' weaving patterns (fig. 4.2). The context, however, was entirely different. Jacobs described the path of a billiard ball on tables of different dimensions and asked the reader to predict in which corner the ball would end up. Eventually this led to a geometric method to determine the greatest common divisor of two numbers, a topic that Gerdes (1988) also developed, but from the point of view of African sand drawings.

My students were interested in the juxtaposition of the two approaches, and they certainly gained a new perspective on the content of mathematics. One young white woman, an artist, generally hated mathematics but was intrigued by this topic and did some original work along these lines. Soon thereafter her father came to school to complain that I was not teaching his daughter mathematics. According to him, I was "just wasting time playing around." He was not the sort of person who could be convinced otherwise, and he had his

Fig. 4.2. Networks drawn in sand by Bakuba children

daughter drop the course. Perhaps he would have reacted more favorably if we had kept the parents better informed about the content and goals of the curriculum.

In general, this infusion of the cultural aspects into the curriculum was extremely successful. African-American students, in particular, found such materials relevant to their lives and backgrounds. One young woman chose to instruct an eighth-grade class in African applications of network theory as her senior year project.

Several social studies teachers began to incorporate African mathematics into their curriculum. One instructor assigned my article on Yoruba (Nigeria) mathematical practices (Zaslavsky 1970) to his world history classes. Several of their students chose to write book reports on *Africa Counts* (Zaslavsky 1979a).

Many mathematical topics lend themselves to the infusion of cultural applications. In the process of designing a rug similar to those woven by Navajo women, students deal with symmetry, geometry, and measurement (Zaslavsky 1990). At the same time they learn to appreciate the degree of mathematical knowledge required to weave complex designs, generally with no pattern and no specific instructions to follow. With this understanding, "women's work" takes on new significance. To mention just one problem, how does the weaver compute the number of rows required to make the border pattern "come out right" when she reaches the far side of the rug?

Mathematics comes alive when students participate in activities that illustrate how mathematical decisions arose from the basic needs of societies. For example, why do people build their homes in certain shapes and sizes and use particular materials? An investigation into styles of building in various cultures is a valuable source of experiences with shapes and sizes, perimeter and area, estimation and approximation, while at the same time it shows the relevance of mathematics to social studies, art, and other subjects (Zaslavsky 1989). Students might consider a tipi, an African mud-and-wattle round house, or an Inuit igloo a "primitive" dwelling compared with an urban apartment house or suburban ranch house. Yet the people who build these homes, or had built them in years past, are using their available materials and technology to the best advantage.

Teachers must be careful that they do not introduce cultural applications as examples of "quaint customs" or "primitive practices." These applications must form an integral part of the mathematics curriculum. They must inspire students to think critically about the reasons for these practices, to dig deeply into the lives and environment of the people involved. It is so easy to trivialize the concept of multicultural education by throwing in a few examples as holidays approach. Better not to do it at all!

The multicultural context is relevant to many aspects of the mathematics curriculum. A discussion of the number words and numeration systems of non-English-speaking peoples may do wonders in raising the self-esteem of students who speak these languages as well as enhancing the understanding of all students. It may come as a surprise that in some languages grouping is by twenties rather than by tens as in English. Games of chance and games of skill and patterns in art and architecture are all sources of learning experiences (Krause 1983; Seattle Public Schools 1984; Zaslavsky 1985, 1987, 1991). Some of the richest contributions may come from students and their families.

In 1985 the distinguished Brazilian mathematician Ubiratan D'Ambrosio founded the International Study Group on Ethnomathematics. The term *ethnomathematics* implies the influence of sociocultural factors on the teaching and learning of mathematics. The prefix *ethno-* encompasses identifiable cultural groups, such as national-tribal societies, labor groups, children of a certain age bracket, and professional classes and includes their jargon, codes, symbols, myths, and even specific ways of reasoning and inferring. There is ample evidence that people in all societies devise their own ways of doing mathematics, independently of their technological level or school learning. Yet these practices are rarely recognized in formal school mathematics. For more information, see Gilmer (1990).

Bringing the world into the mathematics class by introducing both cultural applications and current societal issues does motivate and empower students. Such mathematical content offers wonderful opportunities for project work, cooperative learning, connections with other subject areas, and community involvement. To carry out such a program effectively requires a new approach to curriculum development, teacher education, and assessment processes.

References

Ascher, Marcia. "Graphs in Cultures: A Study in Ethnomathematics." *Historia Mathematica* 15 (1988): 201–27.

Banks, James A. "The Transformative Approach to Curriculum Reform: Issues and Examples." In *Empowerment through Multicultural Education,* edited by Christine E. Sleeter, pp. 125–41. Albany: State University of New York Press, 1991.

Bishop, Alan J. *Mathematical Enculturation.* Dordrecht, Netherlands: Klüwer, 1988.

Brown, John Seeley, Allan Collins, and Paul Duguid. "Situated Cognition and the Culture of Learning." *Educational Researcher* 18 (January/February 1989): 32–42.

Cotton, Anthony. "Anti-racist Mathematics Teaching and the National Curriculum." *Mathematics Teaching,* no.132 (September 1990), pp. 22–26.

Frankenstein, Marilyn. *Relearning Mathematics*. London: Free Association Books, 1989. (United States distributor: Columbia University Press, 136 South Broadway, Irvington, NY 10533).

Gerdes, Paulus. "On Possible Uses of Traditional Angolan Sand Drawings in the Mathematics Classroom." *Educational Studies in Mathematics* 19 (1988): pp. 3–22.

Gilmer, Gloria. "An Ethnomathematical Approach to Curriculum Development." *Newsletter* [of the International Study Group on Ethnomathematics (ISGEm)] 5 (May 1990): 4–6. (Dr. Gilmer is president of the ISGEm and may be contacted at Math-Tech, 9155 North 70th Street, Milwaukee, WI 53223.

Hudson, Brian. "Global and Multicultural Issues." *Mathematics Teaching,* no. 119 (1987), pp. 52–55.

________. "Global Perspectives in the Mathematics Classroom." In *Mathematics, Education, and Society,* edited by Christine Keitel et al., pp. 87–89. Paris: UNESCO, 1989.

Jacobs, Harold R. *Mathematics: A Human Endeavor.* New York: W. H. Freeman & Co., 1970. (Revised 1982).

Janvier, Claude. "Contextualization and Mathematics for All." In *Teaching and Learning Mathematics in the 1990s,* 1990 Yearbook of the National Council of Teachers of Mathematics, edited by Thomas J. Cooney, pp. 183–93. Reston, Va.: The Council, 1990.

Joseph, George G. *The Crest of the Peacock: Non-European Roots of Mathematics.* London: I. B. Tauris, 1991. (Distributed in the United States by St. Martin's Press.)

Krause, Marina C. *Multicultural Mathematics Materials.* Reston, Va.: National Council of Teachers of Mathematics, 1983.

National Council of Teachers of Mathematics. *Curriculum and Evaluation Standards for School Mathematics.* Reston, VA: The Council, 1989.

Research Advisory Committee of the NCTM. "The Mathematics Education of Underserved and Underrepresented Groups: a Continuing Challenge." *Journal for Research in Mathematics Education* 20 (July 1989): pp. 371–75.

Reyes, Laurie Hart, and George M. A. Stanic. "Race, Sex, Socioeconomic Status, and Mathematics." *Journal for Research in Mathematics Education* 19 (January 1988): pp. 26–43.

Schwartz, Richard H. *Mathematics and Global Survival.* Needham Heights, Mass.: Ginn Press, 1989.

Seattle Public Schools, Mathematics Office. *Multicultural Mathematics Posters and Materials.* Reston, Va.: National Council of Teachers of Mathematics, 1984.

Zaslavsky, Claudia. *Africa Counts: Number and Pattern in African Culture.* Brooklyn, N.Y.: Lawrence Hill Books, 1979.

________. "Bringing the World into the Math Class." *Curriculum Review* 24 (January/February 1985): pp. 62–65.

________. "How Do We Spend Our Money?" In *Educating for Global Responsibility,* edited by Betty A. Reardon, pp. 73–79. New York: Teachers College Press, 1988.

________. *Math Comes Alive: Activities from Many Cultures.* Portland, Maine: J. Weston Walch, 1987.

________. "Mathematics of the Yoruba People and of Their Neighbors in Southern Nigeria." *Two Year College Mathematics Journal* 1 (1970): pp. 76–99.

________. "Multicultural Mathematics Education for the Middle Grades." *Arithmetic Teacher* 38 (February 1991): pp. 8–13.

________. "Networks—New York Subways, a Piece of String, and African Traditions." *Arithmetic Teacher* 29 (October 1981): pp. 42–47.

________. "People Who Live in Round Houses." *Arithmetic Teacher* 37 (September 1989): pp. 18–21.

________. "Symmetry and Other Mathematical Concepts in African Life." In *Applications in School Mathematics,* 1979 Yearbook of the National Council of Teachers of Mathematics, edited by Sidney Sharron, pp. 82–97. Reston, Va.: The Council, 1979.

________. "Symmetry in American Folk Art." *Arithmetic Teacher* 38 (September 1990): pp. 6–12.

________. "World Cultures in the Mathematics Class." *For the Learning of Mathematics* 11 (June 1991): pp. 32–36.

Part 2
Changing What Students Learn

The movement for reform in mathematics education calls for changes in the mathematics that students learn. This movement emphasizes such mathematical topics as proportional thinking and geometric reasoning that focus on thinking processes rather than the memorization of mathematics facts.

A number of programs are implementing such changes in mathematical content. The chapters presented in this section represent a small sample of the efforts being directed at changes in the mathematics curriculum. The programs presented all have a "common denominator." In addition to the restructuring of the mathematics content, those programs focus on improving the achievement of students traditionally underrepresented in mathematics.

5

Equation for Success: Project SEED

Sherri Phillips
Hamid Ebrahimi

It has usually been believed that one had to wait until a child was older before introducing him to abstract mathematics. This idea is precisely wrong; it is not that one has to wait, but rather that one cannot afford to wait.

David Page, University of Illinois mathematics professor

Public education in the United States has undergone close scrutiny in the face of falling standardized test scores, unfilled positions in technical fields, and a growing gap between achievement levels of white and minority students. The ongoing crisis in education, particularly in mathematics and the sciences, and consequently in technical and technological fields, has been dramatically and consistently heralded in recent government and private reports.

The net result of the overall education of our children today is unimpressive, since we have failed to prepare large sectors of our population for academic survival. Without substantively improving students' attitudes about themselves and their education, without raising the achievement levels of at-risk students, particularly in mathematics, and without thoroughly restructuring the training of our teachers, any hope of recovery will remain simply a hope.

Many factors contribute to low academic achievement. A chief factor, particularly for disadvantaged students, is their lack of self-confidence. The low expectations of student performance currently held within our learning institutions result in and sustain mediocrity. For students to succeed (and this is especially true for the at-risk sector), they need significant experiences of success in significant areas.

Purpose

The purpose of this paper is to describe Project SEED briefly, to develop its history and philosophy, to demonstrate some of its techniques and methods, to compare its standards with those of the National Council of Teachers of

Mathematics (NCTM), and to share some of the reasons we believe that, in its thirty-year history, it has contributed to the goal of reaching all students with mathematics.

In particular, Project SEED—

1. reaches a diverse body of students with mathematics;
2. believes that this diversity fuels a dynamic and successful learning environment;
3. uses a variety of instructional strategies;
4. draws its curriculum from such advanced topics as algebra, calculus, number theory, and complex variables;
5. is driven by the knowledge that all students are capable of learning higher mathematics;
6. seeks to raise students' expectations of themselves, as well as to positively affect teachers' and parents' expectations of the students.

One Solution

Project SEED, conceived over thirty years ago, is a national mathematics program that serves all students, particularly elementary school students in the at-risk sector. The program, through a carefully developed curriculum and teaching methodology, focuses on improving students' critical-thinking skills and their self-esteem and lets them experience success in higher-level mathematics.

Project SEED seeks to prepare young students, especially those from disadvantaged environments, for success in upper-level courses at the secondary school and university levels and to increase significantly the number of students seeking careers in technological fields. The program combines a fully trained staff of mathematicians and scientists, a rigorous mathematics curriculum, a strategically developed teaching methodology, and an ongoing teacher-training program in order to reach its desired goal: to involve students in an interactive educational environment in which they will increase their critical-thinking skills, enhance their self-esteem, and, most important, become integrally involved in their own education.

A Brief History of Project SEED

In 1963, William F. Johntz, a mathematician and psychologist, began searching for the factors responsible for the deep disparity between opportunities for minorities and nonminorities. Johntz taught lower-level mathematics at Berkeley (California) High School, where his classes were consistently composed of low-income, minority students.

Since the sixties, conventional wisdom has held that low-income children enter kindergarten lacking the verbal skills common to middle-class children. Their academic progress is impeded both by this fact and by the negative self-image that it creates. Johntz reasoned that since

> conceptually-oriented mathematics is seldom a familiar part in any child's environment . . . it is "culture free" [and] the disadvantaged child cannot use it to compare himself unfavorably with others . . . . Success in this subject helps the [educationally] disadvantaged child improve his poor self-image, thereby enabling him to perform better in all subjects, including mathematics. (NCTM 1972, p. 384)

Following his own reasoning, Johntz volunteered his services on his lunch hours and free periods to the Washington Elementary School across the street, teaching algebra and other conceptual mathematics through a Socratic discovery method. Not only did the students learn the concepts, but their basic mathematics skills improved as well. Significantly, they enjoyed and actively engaged in the process of learning.

Over the next few years, Johntz was joined by mathematics faculty and graduates from the University of California at Berkeley, and the project began to grow. State funding in California and Michigan led to the incorporation of Project SEED in 1970 as a nonprofit organization. Since that time, Project SEED has operated programs in numerous other cities and states with local and national support.

Thirty years later, the brainchild of those first days at Washington Elementary School has expanded into a national program, currently reaching students in Berkeley, Oakland, Dallas, Philadelphia, Detroit, and other cities.

Methodology and Techniques

In Project SEED, professional mathematicians and scientists from universities, research corporations, and the community teach abstract, conceptually oriented mathematics to full-sized classes of students as a supplement to their regular curriculum. The instructional methodology is a nonlecture, Socratic group-discovery format designed to permit children to discover mathematical concepts by answering a carefully planned sequence of questions posed by the SEED instructor. Higher-order mathematical topics are selected to reinforce and improve the students' critical-thinking and computational skills and to help equip them for success in higher mathematics courses, both at the secondary school and university levels. Project SEED teaches regular school classes rather than specially selected groups of students. Although Project SEED concentrates its efforts on the elementary school level, it has recently increased its middle school emphasis and provides curriculum and instruction for all levels from elementary to graduate school.

Project SEED's long-range goal is to increase the number of minority and educationally disadvantaged youth majoring in, and pursuing careers in, mathematics and related fields. Project SEED attains this objective by raising the achievement level and consequently improving the self-concept of children by providing them with success in a high-status, abstract subject. It is important that this subject not be associated with past failure. The direct remediation that characterizes most compensatory education programs often fails because it tends to derogate the child by concentrating on the areas in which he or she already has had negative experiences. Project SEED embeds arithmetic in the SEED curriculum (i.e., high school and college algebra), which is new and therefore free of connotations of past failure. Mathematics is a universal language that eliminates preexisting cultural fences between students.

Why Start at Such a Young Age?

Students in the early years (Project SEED begins teaching at the early elementary school level) are still intellectually open and curious. It is in these grades, where crucial attitudes toward school and learning form, that it is critical to present academic inquiry at its best. Programs where students experience an enjoyment of learning—programs that instill an outlook that education is fun and intellectually engaging—are vitally needed at the elementary school level now and on a continuing basis.

By the time students reach high school, the overwhelming majority are already mathematically scarred; not only do they dislike mathematics, but for the most part they are unqualified to enter algebra and other higher mathematics courses. One possible reason for this educational devastation is that elementary school teachers are generalists and, through no fault of their own, often have not had the advanced study of mathematics that precipitates fluid verbalization of the material and can frame children's learning in a larger mathematical picture. For this reason, Project SEED uses trained specialists.

Involving the Students

Our experience in Project SEED leads us to believe that even children who are failing in school can exhibit enormous competence and joy in learning high school and college level algebra when their classes are taught by a trained specialist—someone who loves mathematics and understands it in depth. Students' mastery of the prestigious subject can improve their attitudes and consequently their performances in nonmathematical areas.

The Socratic method, when used by trained instructors, solicits consistent student participation. Project SEED's carefully developed techniques draw participation from the class as a whole and involve those individuals who are normally shy, confused, or unruly. Project SEED instructors are trained to ask a series of questions designed to aid the children in discovering mathematical concepts and then to explore every student's response for his or her mathematical thinking.

Sequence of Questions for First Day—Grade 5

Following is a very brief example of questions asked on a typical first day of fifth grade, based on the introduction of the concept of exponents. "E" is used as the operation, since students are accustomed to seeing a symbol for binary operations. The letter *E* also adds an element of excitement in the classroom as students try to determine what this new operation could mean. Standard power notation is reinforced later in the curriculum through a very simple transition.

The following operations, developed by the students, are written on the chalkboard:

Addition	6	+	2	=	8
Subtraction	6	−	2	=	4
Multiplication	6	x	2	=	12
Division	6	÷	2	=	3
	6 E 2 =				

Note that the numbers have been chosen with care: 6 and 2 have solutions in the set of natural numbers for all the problems; they are small enough to require little calculation, thereby allowing students of all ability levels to participate; and the numbers are different, reducing possible confusion.

Raise your hand if you can use a 6 and a 2 together to get an answer other than what we already have on the board.

Here, the specialist takes each conjecture and carefully listens to explanations of how the answers were deduced. This is a prime opportunity to set the entire tone of the classroom by praising logical thinking, by encouraging verbalization, by prompting students to explore one another's thinking, by reinforcing the use of agreement/disagreement hand signals to show interest in every opinion, and by establishing a rapport with the students. (It may be fun to try to figure out some of the more typical solutions that are given for 6 E 2: 62, 26, 652, 60, 13, 27. All these answers are produced by students through a logical thinking process.) When the answer of 36 comes up, which it invariably does, it receives high praise.

Excellent! On the count of two, everybody read the complete sentence together.

The specialist avoids getting an immediate explanation in order to allow others the joy of the discovery.

Raise your hand if you can use 7 and 2 in the same way to get an answer.

Again, conjectures are given, but fewer this time. As soon as 49 is stated as a conjecture, it is placed on the board, and 8 E 2 is posed. Now, students are literally begging to be allowed to explain the system as more and more catch on.

I wonder how many people already have the system for our new operation E. Raise your hand if you have the answer for 9 E 2. Show me by your signals if you agree with that answer.

What problem did you work to get 49? What's the product of 7 x 7, everyone? That seems like it might work. Let's see if the system works on the others. What problem did you work to get the 81? What's 9 x 9, everyone?

The other problems are stated and placed on the board.

I see how you're using the 6, the 7, the 8, and the 9, but who can tell me how you used the 2? Would someone else like to explain in your own words how you used the 2?

High praise takes place here for the verbalization of concepts and ideas. A fair amount of chorus reading and chorus answers also occur, reinforcing the concept of teamwork, focusing the class on the mathematics, and allowing group participation.

The board now looks like this:

Operations

Addition	6	+	2	=	8
Subtraction	6	−	2	=	4
Multiplication	6	x	2	=	12
Division	6	÷	2	=	3
	6 E 2	=	6	x	6
	7 E 2	=	7	x	7
	8 E 2	=	8	x	8
	9 E 2	=	9	x	9

If that is our system, then show me with your signals—do we still have a true mathematical sentence if I put another factor in the factor form? [No.] Who can tell me what I'd have to do now to make it true? [Change the 2 to a 3.]

This is especially important. As students explain how the exponent 2 is being used, it is easy for them to see what should change when the specialist puts an additional factor in the factor form. Again, student responses are reinforced as being valuable. Every signal—agreement, disagreement, "I don't know"—is praised highly. The word *exponent* is used in context as the specialist erases the 2 in 6 E 2 and replaces it with a 3. These contextual clues need only be used a few times before students adopt them and begin using them in context. As soon as a student uses the vocabulary, usually in the course of an explanation, the entire class is asked if they heard that word. They are asked to repeat it and to point to where it might be represented numerically on the board. Quick drills ensue, using the vocabulary word repeatedly. Most important, the concept of exponents rapidly becomes clear as the specialist uses an erase-and-change technique to substitute different exponents and factor forms in the statements.

If I change the 3 to a 5, who can tell me what I'd have to change in the factor form on the right-hand side? Would I use the chalk or the eraser to make it true?

Interestingly, students have no problem with new vocabulary as long as contextual clues are given. It is adults that will sometimes interrupt and say, "I don't know what a factor form is!" Several questions of the same type follow, with frequent opportunities for explanations by students. "Exponent" and "factor form" rapidly become familiar to the student, both conceptually and as vocabulary.

Who can tell me why we use 6s in this factor form?

[6 E 3 = 6 x 6 x 6]

Students will point out the 6 in the exponential form. This provides the transition to begin exploring the role of the base.

Do we still have a true math sentence if I change the factors to 4s? Point to something I'd have to change. Stop my eraser when I get there. What should the base be, everyone? If I change the base to a 5, what should all the factors be? If I change the base to a 10? If I make all the factors 2s, what do I need to change? Should I change the exponent? The base?

The board will be developed as follows, which is the setup for looking at the additive law for exponents. The goal in the first few class periods is to have the students fluent in the basic concepts of exponentiation, using the vocabulary in context, and able to verbalize the role of the base, the operation, the exponent, quantities, and factor forms. Enthusiastic discussion ensues about the problem in the middle of the chalkboard, through which the students will discover the additive law for exponents. They will need only the tools that have been introduced up to this point.

2 E 3 = 2 x 2 x 2
2 E 4 = 2 x 2 x 2 x 2
2 E 5 = 2 x 2 x 2 x 2 x 2 (2 E 3) x (2 E 4) =
2 E 6 = 2 x 2 x 2 x 2 x 2 x 2
2 E 7 = 2 x 2 x 2 x 2 x 2 x 2 x 2

The previous set of questions is only the starting point as the fifth-grade students search for other laws of exponentiation (multiplicative law for exponents, multiplicative law for bases, subtractive law for exponents, etc.) and move on to discover inverse operations, logarithms, summations, and other advanced concepts. This will require no lecture from the specialist but necessitates careful lesson planning, constant awareness of the students, consistent use of SEED techniques, and most important, the desire and ability to take advantage of the spark when it occurs. Unfortunately, the writers have been severely limited here; there is no way to transcribe the depth of the dialogue that takes place among the students or the enormous excitement that the students feel when discussing mathematics. The questions provide the impetus; the students provide the joy.

With the introduction of the problem above [(2 E 3) x (2 E 4)], the reader can see the richness of the curriculum. As students give their conjectures and discuss the logic behind each idea, all the following items are reinforced: exponentiation concepts, proof structures, grouping, two different operations in one equation, chain multiplication and associated shortcuts, patterns, basic mathematics concepts (baseline material), verbalization skills, logical and critical-thinking skills, and such social skills as listening, respectfulness, and patience. In the course of each explanation, students will use previously learned material while stretching their minds to reason through new concepts. This provides ample opportunity for every student, regardless of skill level, to contribute to the class.

The Value of Each Student's Response

In the firm belief that respect for the students leads to their heightened participation, SEED specialists investigate the ideas of the students, right or wrong, in the classroom. Each explanation by a student is afforded complete attention, under the assumption that it is based on a valid and careful thought process. Often, the answers given by the students have been reached through creative, sometimes profound logic. An actual example of the imaginative, innovative thinking of children as reported in *Think* magazine, is given in figure 5.1.

The intellectual risk-taking that such a format encourages is invaluable, both to the diversity of the classroom and to the academic future of the students. Once the students realize that they can express their ideas and that their thoughts are both intelligent and original, many of the mental and emotional boundaries, both self-imposed and external, are erased.

An Important Indirect Approach: Letting the Student Talk

Project SEED encourages students not only to investigate what is, but even more important, to search out what "is not." Some of the most valuable discourses take place around this concept. Why, even though a particular answer is logical and seems to fit, is it not the best possible conclusion? The problem expressed in the equation $2^{-1} = \alpha$ eventually is solved by the students, with general agreement that $\alpha = 1/2$. Why, however, is the answer not -2, or 1? This extra step is vital in firmly rooting the concepts in the students' minds.

The floor is always open for students' conjectures, for their explanations of carefully thought-out theories. As the students discover age-old concepts, they gain an ownership of the mathematics through their discovery—that knowledge is now theirs.

Robert's Cartesian Triumph

A SEED class running in high gear excites almost anyone who participates or happens to be watching. Here is a typical mathematical discovery made by a third grader.

In the early days of the program Bill Johntz, who founded it, had spent two or three periods introducing a Berkeley elementary school class to the notion of graphing. In the mathematical terms he freely employed, they had learned that an "ordered pair" of numbers can be mapped as a point in 2-dimensional space. For example, when he wrote (2, 3) on the blackboard and asked: "Who wants to show this point?" every child waved one hand eagerly. He picked a volunteer, who literally skipped to the board, drew a large cross, and starting from the center marked off two steps to the right and three steps upward. The rest of the children shot up both hands, a signal indicating they had the same answer written on paper at their desks. This, Johntz considered, was pretty good thinking by a class that were generally subpar in writing and arithmetic.

Then, without warning, he threw the children a curve ball. On the board he wrote (2, 3, 5) and asked: "What do you suppose this could mean?"

For about three minutes the children stewed and scribbled on their note paper. Her and there, two conferred with each other. Then one little boy tentatively raised his right hand. "All right, Robert," said Johntz, "suppose you tell us what it means to you."

Robert explained that he imagined a glass tube marked off in steps like a thermometer. "You hold it so it sticks straight out from the board. Then you slide it along so the end is on point (2, 3). Then you walk out five steps on the tube and that is point (2, 3, 5)."

"What do you think of Robert's idea?" Johntz asked. A few children raised both hands in agreement. A few others crossed their arms, signaling they thought Robert was wrong. And then the class began a debate on the merits of Robert's new theory.

Of course, Robert was basically right. Although it was hardly necessary to imagine a glass tube, he had just the same made an important invention, which mathematicians call "3-dimensional Cartesian coordinates." It is the usual way of identifying every point in space by a unique trio of numbers. By translating geometry into algebra it enables physicists and engineers to calculate such things as rocket trajectories or stresses on a steel girder.

Everyone present enjoyed Robert's triumph for a variety of reasons. The regular teacher, sitting by while Johntz taught algebra, was thrilled to see her class grapple successfully with a tough mathematical concept. Two parents, watching from the back of the room, were stunned to hear their own children volleying ideas they didn't comprehend in language they hardly understood.

And Johntz himself, an ardent admirer of children, was excited because Robert had discovered a mathematical idea on his own. For such triumphs a growing number of mathematicians are devoting part of their careers to teaching elementary schoolchildren.

Fig. 5.1. Reprinted by permission from *Think* Magazine. Copyright 1970. International Business Machines Corporation.

The specialist may make deliberate errors in the classroom as the students revel in their newfound knowledge. The errors are used both to reinforce the students' knowledge and confidence and to encourage them to articulate the concepts behind their answers. The specialist will offer variations of the same problem from different vantages, allowing the students to increase the depth of their understanding through their recognition of a concept presented in different guises.

A Silent Language for Continual Participation

One of the trademarks of Project SEED over its twenty-seven-year history has been its silent language, used in addition to a constant verbal exchange and developed to allow all students to participate at once, providing the specialist with a constant measure of the understanding, the confusion, and the level of participation of each member of the class. Any observer of a SEED class will notice the students' hand signals immediately: arms raised straight in the air signal agreement and understanding, arms rapidly crossed back and forth in front of the body signal disagreement, and a variety of other signals indicate partial agreement, confusion, and questions.

The students immediately begin using the signals to express themselves, extending the "language" to outside the class as well. Recently, one class gathered for a field trip to a local museum. Questioning the students in a rapid-fire manner as he sought their reactions to the paintings, the museum director was confused by a sudden flurry of arms waving in the air. The students had automatically responded to his questions with their SEED signals as they tried to remain silent but assert their opinions.

Movement of the Body, the Voice, and the Material

Another important aspect of SEED instruction is a constant spatial awareness of the classroom. The specialist moves about the room, stopping by potential trouble spots, energizing the students through a more active presence in the classroom, casually pausing by this or that shy student for a careful moment of attention. The simple act of moving about the classroom causes students to remain alert, decreasing the possibility of a student's daydreaming or playing with friends. Additionally, students are allowed only paper and pencil on their desks, which centers their attention on the specialist and other students.

Specialists also maintain students' interest by constantly modifying the pace of the class, through increased tempo and volume of the voice, through a change of pace of the material, and through the use of different techniques on a nonpatterned basis. Using these techniques, the specialist can effectively address various types of students, from disruptive, aggressive students to shy, retreating students. Project SEED has had great success in addressing the special education students as well.

The net result of this combination of methodology, techniques, and curriculum is a highly interactive class in which students and specialist participate in an intellectual discourse, fostered by intense curiosity and mutual respect.

The Crucial Importance of Effectively Training Teachers

Each of the benefits above derives from a comprehensive training program. As an absolute minimum, Project SEED requires the following of every specialist:

1. Each specialist must have a strong background in advanced mathematics (a minimum of a degree in mathematics or its equivalent).
2. Regardless of previous teaching experience, each specialist must undergo a multidimensional training program.
3. Every specialist has a schedule in which he or she observes and critiques others' classes and is observed and critiqued by peers on a frequent basis.
4. Specialists must attend workshops in advanced mathematics, methodology, classroom psychology, and other related fields every week.
5. Specialists must become capable of continually revising and extending curriculum and methodology.

A superb program requires a dedicated, highly qualified staff. Staffing, which is probably one of the more difficult hurdles for agencies that need professionals with expertise in technical fields, is carefully undertaken by Project SEED.

Prospective SEED staff are screened first through interviews, then through an intensive two-week initial training and selection program. They attend classes on curriculum, methodology, classroom techniques, and various mathematical topics designed to enrich their knowledge and give them more depth in the classroom. They must complete a series of dry runs in front of the other specialists in which they demonstrate the curriculum and methods that they are learning. Only at the conclusion of this initial training and selection process are these applicants considered for trainee positions.

Complete training never ends in Project SEED. The workshops continue throughout the entire school year and into the summer. Advanced mathematics workshops are given periodically to enhance and reinforce the concepts underlying the curriculum, giving the specialists themselves the opportunity to learn by discovery. As one specialist noted, "In order to be a good teacher, one must remain a good student."

The self-evaluation aspect of Project SEED is one of its unique traits. In addition to teaching their classes, specialists observe and critique one another's classes on a weekly basis. Although this is sometimes intimidating to new specialists, they rapidly learn to depend on the regular feedback, from which they can make instant adjustments to their classes.

Training must remain ongoing in any program that desires to keep its staff fresh, involved, and aware of the dynamics of the classroom. Group training not only serves the purposes of peer evaluation but also builds a sense of unity and teamwork among the staff. Continual reinforcement and review are necessary to maintain a program's integrity and its quality.

Leaving a Legacy in the Classroom

The regular classroom teacher is always present as a participant and an observer when the SEED mathematician is working with the class. It is the intention of the SEED program that traditionally low expectations of students will be significantly heightened as teachers see the children (including the "low achievers") consistently understanding and manipulating difficult mathematical concepts.

The teachers' training available in Project SEED differs from traditional in-service programs because the SEED specialist demonstrates the new techniques and attitudes daily in the regular teachers own classroom. In addition to in-class modeling, Project SEED offers school districts and universities special in-service programs focusing on mathematics and methodology. Staff developments, seminars, and workshops are also offered. Evaluations of SEED in more than 200 classrooms have consistently demonstrated a very positive impact on the teachers in whose classes the program was operating.

The Curriculum

Young children can comprehend complex concepts; their minds are open and quick, and they have not had time to form biases. Often, students will have an understanding of concepts that elude adults; for example, the concepts of infinity and limits are often easier for a child to grasp, facilitating the learning of such topics as infinite series, as in the following equation:

$$\lim_{n \to \infty} \sum_{\beta = 1}^{n} (4 \times 5^{-\beta}) = \square$$

This equation, which is developed in fifth or sixth grade after concepts of positive and negative exponents, summations, limits, and adding fractions with unlike denominators, leads to discussions about partial sums with respect to limit proofs. We have seen students not only quickly grasp the concept but become excited by the idea of being able to add an infinite series of expressions and still arrive at a finite number for an answer.

The majority of the curriculum has been developed around concepts from high school algebra and university-level mathematics. Algebra is the gateway to higher mathematics and, as it is taught in Project SEED, serves as a pump rather than a filter, leading more students to higher mathematics rather than discouraging them from the start. The critical-thinking skills that students gain in the classroom are essential to their success in higher-level courses as well as in subsequent careers.

All SEED curriculum is advanced in a spiral, in which the specialist reviews concepts from earlier classes, works with the current material, and introduces new concepts to lead into more advanced material. This spiraling path is extremely important, not only in cementing the mathematical concepts in the minds of the students but as a method of involving every student in the classroom. The more capable students are able to advance quickly while the slower students have the advantage of seeing the concepts developed repeatedly through different topics in the curriculum. For the entire class to succeed, there must be a morsel for every student to taste.

The concept of variables is introduced almost immediately in the SEED class (students rapidly become familiar with the Greek alphabet), facilitating the generalization of complex concepts. For instance, students generalize the additive law for exponents, which they commonly call ALFE.

$$\alpha^{\beta} \times \alpha^{\gamma} = \alpha^{(\beta+\gamma)}$$

This equation leads to the discussion of specific problems, such as the following:

$$2^5 \times 2^{-2} = 2^3$$
$$32 \times \Delta \ = 8$$

By using this preceding problem, which has usually been developed and generalized by the second week of a fourth- or fifth-grade class following familiarization with positive exponents, the specialist is able to introduce the idea of negative exponents while reinforcing the concepts of additive and multiplicative inverses. Basic computational skills are embedded in the material as well.

Since the students have already learned that the numerical value of 2^5 is 32 and of 2^3 is 8, the only possible answer for 2^{-2} is 1/4, which the students discover through a series of their own conjectures. As seen in the equation above, this is developed through a vertical proof structure. Proof structures are an essential element of the SEED class, enhancing the logical and conceptual reasoning of the students.

Discussions of ALFE lead students to develop conjectures of such abstract notions as 2^0 and 0^0. The depth of these discussions often reaches a level of sophistication paralleling that of an enthusiastic group of graduate students.

As the mathematics is developed, so is the vocabulary of the students. The correct pronunciation and spelling of mathematical terminology is important, as the students proudly use words that "only real mathematicians know." In the course of a SEED class, the students become familiar with such words and phrases as *exponentiation, logarithm, function, distributive law, associative law, summation, additive inverse, multiplicative inverse, argument, index of summation, conjecture, inference, hypothesis,* and numerous other terms.

The curriculum of Project SEED is varied. Students sample topics ranging from group and field properties to limits to elementary complex variables. By letting the students experience success in a difficult and prestigious subject, Project SEED increases their self-esteem, improves their mathematical skills, and prepares them for higher-level mathematics and careers. Ultimately, the curriculum is used in tandem with the methodology to achieve the overall goals of Project SEED.

The Bottom Line: Results

Those aspects of Project SEED that can be evaluated by existing and reliable tools, such as those that assess student basic skills achievement or attitudes of teachers, parents, and principals, have been formally evaluated throughout its years. SEED has been evaluated on such nationally normed tests as the CTBS and the ITBS and through surveys of students, teachers, and principals.

Early evaluations of Project SEED revealed not only that children receiving SEED instruction were able to perform abstract, conceptually oriented mathematics but that their arithmetic computational skills increased enormously. Evaluations over a five-year period (1975–80) conducted by Educational Planners and Evaluators found that SEED students nationwide averaged approximately two months' growth in arithmetic for each month they participated in the program. After five years of studying the program, the EPE evaluators stated in their evaluation that "Project SEED unquestionably fosters improved arithmetic skills in the vast majority of participating students, and that these five national evaluations provide overwhelming evidence of the ability of SEED to stimulate mathematical thinking in young children, which enhances both their conceptual and computational skills."

The most in-depth analysis was recently concluded (November 1990) by the Research and Evaluation Department of the Dallas Independent School District. It conducted a multiyear longitudinal study of the effects of SEED instruction on students' mathematics achievements and attitudes, using experimental and comparison groups of students matched by sex, ethnicity, race, socioeconomic status, and pre-SEED instruction evaluation. The major results of that study are as follows.

- SEED instruction positively affects students' attitudes toward mathematics. The study showed some significant increases in the enrollment of SEED students over non-SEED students' in advanced mathematics courses at the middle and early high school levels.

- SEED instruction positively affects students' overall performance. Therate of grade repetition in the study was significantly lower for SEED students than for non-SEED students.
- SEED has an immediate and positive effect on students' mathematics scores as measured by the ITBS. Students who had taken only one semester of SEED scored higher on all mathematics sections of the ITBS as compared to non-SEED students.
- SEED has a cumulative effect one students' mathematics scores. Not only did SEED students outperform non-SEED students on the ITBS after only one semester of SEED instruction, but the margin between their scores increased significantly for each additional semester of SEED instruction.

Unfortunately, most evaluation methods do not have a means of accurately measuring the psychological changes that appear to occur among students in SEED classes—intense eagerness, joyful participation, and depth of thought on the part of the students. Learning to love education is a separate issue from education itself; loving to learn is the most solid foundation that any child can have, both in academia and in life.

The Standards of Project SEED and NCTM

NCTM's *Curriculum and Evaluation Standards for School Mathematics* (1989) emphasizes mathematical communication and higher-order thinking skills. According to NCTM, students, while developing these skills, should (1) learn to value mathematics, (2) become confident in their ability to do mathematics, (3) become mathematical problem solvers of open-ended problems, (4) learn to communicate mathematically, and (5) learn to reason mathematically, using inductive and deductive logic. Under the algebra standards, NCTM encourages the development of the understanding of expressions and equations involving variables through the use of a variety of methods.

All these goals are consistent with the goals of Project SEED. Since its inception, Project SEED has developed an effective Socratic discovery methodology and a curriculum based on higher mathematics beginning with algebra that—

1. treats mathematics as a prestigious and valuable subject;
2. provides students with numerous "success" experiences designed to increase their self-esteem and boost their confidence in their mathematical abilities;
3. allows the students to explore open-ended mathematical questions while "discovering" the underlying concepts together;

4. encourages students to participate in intellectual discourse, freely verbalizing their own theories and ideas;
5. develops students' critical-thinking and higher-mathematics skills while reinforcing basic arithmetic skills.

Conclusion

Mathematics is the key to opportunity.... For students, it opens doors to careers.... Jobs that contribute to this world economy require workers who are mentally fitworkers who are prepared to absorb new ideas, to adapt to change, to cope with ambiguity, to perceive patterns, and to solve unconventional problems.

—*Everybody Counts,* National Research Council

The educational needs of our children are great. Project SEED, like other programs, has attempted to meet some of these needs through the strategic development of curriculum and methodology. Project SEED's curriculum is mastered by students from disadvantaged backgrounds in an environment of intellectual discovery and discourse and is guided, in a nonlecture format, by the SEED mathematics specialist.

By providing "success" experiences in higher mathematics, Project SEED develops students' critical-thinking skills, improves their self-concept, and prepares them for success in later life. Project SEED seeks to improve the quality of life for students by instilling in them a love of the pursuit of knowledge and a confidence in their own capabilities and intelligence.

References

National Council of Teachers of Mathematics. *Curriculum and Evaluation Standards for School Mathematics.* Reston, Va.: The Council, 1989.

_______. *The Slow Learner in Mathematics.* Thirty-fifth Yearbook of the National Council of Teachers of Mathematics, pp. 383–87. Reston, Va.: The Council, 1972.

6

The Mathematics and Science Microcomputer Project: A Model for Success

Howard Johnson
Joseph Leonard

Science education must be tailored to the needs of all students, not merely the small minority who go on to full time careers in science. A major reordering of priorities and a redefinition of goals will be required in order to serve the interests of minority groups, women, and the handicapped, as well as all those students destined to be voters, parents, and consumers in our increasingly technological society.

Richard F. Brinckerhoff
and Robert E. Yager
The Second Exeter Conference, 1985

Alarming numbers of young Americans are ill-equipped to work in, contribute to, profit from and enjoy an increasingly technological society. Far too many emerge from the nation's elementary and secondary schools with an inadequate grounding in mathematics, science and technology" (National Science Board 1983). The most ill-prepared are minority students who tend to take fewer courses in mathematics, science, and technology than nonminority students. What is often taken as minority children's failure to learn can just as easily be seen as the school's failure to teach them. Today, African Americans and Hispanics compose about 25 percent of the (K–12) school population, and by the turn of the century, they will constitute 47 percent. In approximately twenty-three of the twenty-five largest school systems in the United States, minority groups are predominant.

Nationally, research continues to demonstrate a disturbing trend on the part of minorities to forego the pursuit of a college education. This, coupled with the tendency on the part of many higher education institutions to augment their mandate of educational opportunity with a commensurate reemphasis on academic selectivity, results in an ever-shrinking pool of eligible African American, American Indian, and Puerto Rican/Hispanic applicants.

It has been established that minority students consistently score below the national norms on standardized mathematics tests and do not enroll in advanced mathematics classes at the high school level (Carpenter et al. 1980; Peng, Fetters, and Kolstad 1981). For example, the data in *High School and Beyond* (Peng, Fetters, and Kolstad 1981) show higher sophomore-to-senior dropout rates for African Americans and Hispanics than for whites—16.8, 18.7, and 12.2 percent respectively. This is probably a conservative estimation, since many minority students leave school before the sophomore year. This is noteworthy because the students dropping out probably are the students having the most difficulties academically. However, research on attitudes indicates that minority students enjoy mathematics, find it interesting, have little mathematics anxiety, and desire to take more mathematics (Becker 1981; Matthews 1984; Nelson 1978). These findings appear to stand in stark contrast to the low achievement scores of many of these students.

Our main purpose was to address this contradiction. We wished to couple positive mathematical and scientific experiences with the students' innate liking for mathematics, thereby increasing the likelihood of their electing to study more mathematics and science during their school career. We sought to do this by creating a learning environment in which students could be taught skills of systematic thinking in the domain of mathematical problem solving. By systematic thinking, we mean thinking that encompasses the whole range of concepts dealing with well-defined processes, including the structure of data acted on as well as the structure of the sequence of operations. This is a requirement for success in problem solving. This type of thinking involves an ability to formulate a problem in usable terms, to establish a plan, and to develop and alter tactics as work progresses.

Papert (1982) suggests that learning to program is a useful way to develop thinking skills, including those of systematic thinking. Others have found that computer programming is a good way to develop appropriate problem solving heuristics (Wells 1981; Blubaugh 1984). At present, however, any comprehensive evaluation of this claim is not possible; the appropriate programming or computer environments are still under development, and students might need to spend several years in such environments before significant effects on their thinking become apparent. Nevertheless, it is generally believed that experience in programming can develop thinking skills to some degree.

The Program

The Syracuse Hill Educational Consortium (consisting of the State University of New York College of Environmental Science and Forestry, the State University of New York Health Science Center at Syracuse, and Syracuse University) in partnership with the six middle schools of the Syracuse City School District conducted a Mathematics and Science Microcomputer Project for 216 minority or economically disadvantaged students during the summers of 1987, 1988, 1989, and 1990.

The program's participants were drawn from the six Syracuse middle schools. Sixty-nine percent of the participants were eighth graders; the remaining 31 percent were seventh graders. There was a culturally rich mix of ethnic backgrounds, which included 29 percent white, 62 percent African American, and 9 percent other. The group was 59 percent female and 41 percent male (see table 6.1)

Table 6.1
Demographic Distribution of Students

	1987		1988		1989		1990		TOTAL	
MALE										
GRADE	7	8	7	8	7	8	7	8	7	8
Ethnicity										
African American	4	8	1	9	6	9	10	8	21	34
Hispanic	0	1	0	0	1	0	0	1	1	2
Native American	0	0	0	0	0	0	0	0	0	0
White	1	7	1	9	3	7	1	1	6	24
Other	0	0	0	1	0	0	0	0	0	1
Total by grade	5	16	2	19	10	16	11	10	28	61
Total male	21		21		26		21		89	
FEMALE										
GRADE	7	8	7	8	7	8	7	8	7	8
Ethnicity										
African American	0	16	2	16	4	14	21	6	27	52
Hispanic	0	0	0	0	1	3	1	1	2	4
Native American	0	1	0	0	0	1	1	0	1	2
White	1	14	4	9	1	3	1	0	7	26
Other	0	0	1	4	0	1	0	0	1	5
Total by Grade	1	31	7	29	6	22	24	7	38	89
Total Female	32		36		28		31		127	
ALL STUDENTS										
GRADE	7	8	7	8	7	8	7	8	7	8
Ethnicity										
African American	4	24	3	25	10	23	31	14	48	86
Hispanic	0	1	0	0	2	3	1	2	3	6
Native American	0	1	0	0	0	1	1	0	1	2
White	2	21	5	18	4	10	2	1	13	50
Other	0	0	1	5	0	1	0	0	1	6
Total by grade	6	47	9	48	16	38	35	17	66	150
Total all students	53		57		54		52		216	

The project was conducted on the campus of Syracuse University. Students participated in a sixteen-day, four-week, summer nonresidential program that included (1) a science component; (2) a mathematics/computer component emphasizing programming skills in Logo and associated mathematical skills; and (3) a career awareness component emphasizing the many career possibilities in mathematics- and science-related professions.

The teachers were drawn from the Syracuse area. Secondary school mathematics teachers with computer interest and experience planned and taught the program. Graduate science teaching students led the science component, and undergraduate education majors were the teaching assistants. The computer laboratory was in a large college mathematics classroom with Apple II computers borrowed from the Syracuse City School District. The science laboratories were in the university's chemistry and physics buildings. Instructional resources came from Syracuse University's Mathematics Education Department's software and materials collection, and the teachers also brought materials from their own schools with them.

The purpose of the Mathematics and Science Microcomputer Project was to provide positive experiences in mathematics and science. Students were introduced to new, unfamiliar problems, which they explored using appropriate problem-solving strategies and skills. They learned to analyze a problem, to generate sound conjectures, to devise and carry out plans for verifying the conjectures, and to form the habit of making a critical review of results.

To achieve this end, computer-interactive lessons focusing on systematic thinking were used. The computer was used for the reasons identified below.

1. Systematic thinking is inherently dynamic and thus seems appropriately taught using a dynamic instrument like the microcomputer.
2. Computer simulation allows the machine to do vast amounts of work, thus allowing the learner to concentrate on strategic aspects without getting bogged down in tactical matters.
3. The computer is fascinating to students. This, coupled with the fact that most minority students have not developed a fixed frame of reference with which to judge success or failure when working with these devices, presents an opportunity to study the relationship of learning, motivation, and attribution of success and failure in problem solving. A computer-based program can set higher expectations for students.
4. Learning becomes student centered. Students interact with the computer and each other using the teacher as a resource.

Since we had a full-day program, we needed varied and wide-ranging activities. A good mathematics activity (or problem) would require the student to use synthesis, would be generalizable, would lead to a variety of solutions, and would involve a mathematical skill. We also wanted the students to have a vision of the technologybased future in which they will work and go to school.

LogoWriter was used for most computer activities. Calculator laboratories and problem solving-software were used as separate units or to introduce skills needed with Logo problems. Science activities used the entire university campus, keeping the students moving and active in the rich arena of architectural structures, medical facilities, and technology labs.

Middle school students came to the summer program with a variety of computer experiences. Some had used Logo, but few had been taught word-processing skills. Word-processing exercises were used as the hands-on introduction to the computer and its operating systems. LogoWriter was chosen because it was suitable as a word-processing program for this age group.

By first learning word processing and disk operations, (such as this insert-and-delete exercise: "Correct the following: Jeanne had an acre in her side"), the students had less frustration when they started programming. Writing exercises throughout the program maintained word-processing skills. The students kept individual journals of their activities on their Logo disks. One group of students even produced a program newsletter.

Students needed to know they were in control of the computer and its software. Taking time to cover basic computer terms gave them a working vocabulary. Naming the operating skills made the use of Logo and other software easier for students and teachers and gave all participants a way to express ideas and ask questions.

Computer hardware and the disk operating system (DOS) were introduced to the students by comparing the computer and its peripheral equipment to an audio stereo system. Basic terms such as *hardware, input, output, play, record, save*, and *load* were introduced. An understanding of the computer and its memory were developed with the students from the comparison of working on a calculator problem. The central processing unit was the calculator, the memory was the scrap paper, and the program was the algorithm the students were following.

Logo mathematics problems for middle school students were designed to fit between the concrete manipulatives of the elementary grades and the symbolic manipulations of high school. The graphics problems used geometric transformations, spatial relationships, estimation, and logical sequencing. Logo primitives were quickly learned by this age group, and writing procedures (programs) started with the first or second lesson. Three exercises were used to introduce or review the primitives and write the first procedures: (1) Moving a turtle (cursor) into a square drawn on the computer screen started the cursor movement and could be written up as a procedure (see fig. 6.1). (2)

patterns done first on graph paper and then with Logo reviewed the primitives; and (3) regular polygons, spinning polygons, and printing exercises introduced the repeat function. Students were quickly writing simple structured procedures as a result.

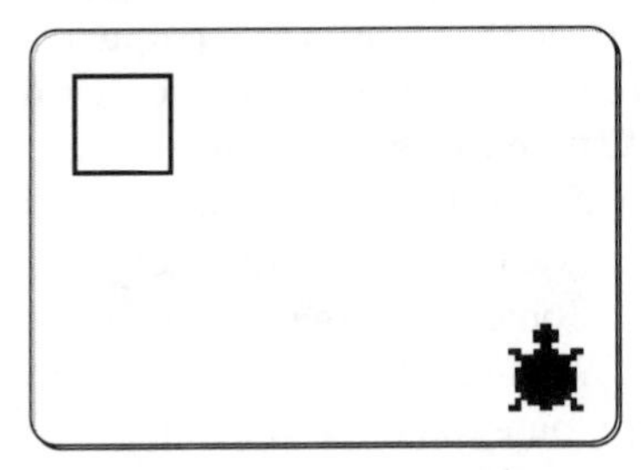

Write a procedure TO PARK the turtle in the garage.

Fig. 6.1. An introductory Logo activity

Transformational geometry is a major thread in the New York State mathematics curriculum. It begins in the intermediate grades with slides, glides, and flips and continues through high school with transformations on graphs of functions. Beginning Logo projects were transformation problems (see figs. 6.2 and 6.3). Rotations were difficult for some students. Therefore, the Factory software from Sunburst was used for a concrete example of rotation and showed that more than one logical sequence can solve a problem. Reflections were started with a pencil and the plastic Mira, which reflects a paper image. Some students were able to write reflective Logo procedures after working with the Mira. Initially, students worked individually on Logo projects to develop their programming skills. They became able to plan and revise strategies with and without teacher help.

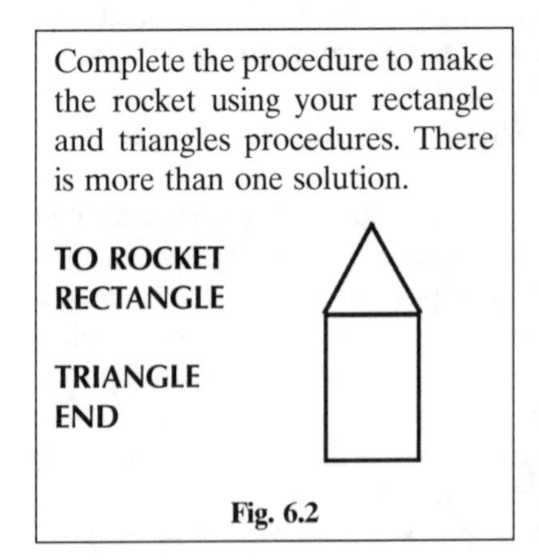

Fig. 6.2

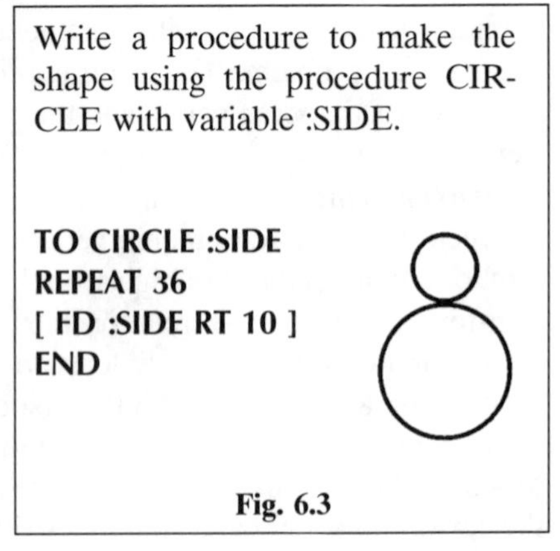

Fig. 6.3

The *Curriculum and Evaluation Standards for School Mathematics,* issued by the National Council of Teachers of Mathematics in 1989, suggests that students "draw on their own interest and background in a free and open environment" (p. 68). Student-initiated Logo projects are done at the end of the program, either in groups or individually. The projects were large, structured graphics procedures, such as a detailed object like a truck or a Logo movie.

Movies made with LogoWriter brought many students back to the computer lab on their own time and ranged from dancing triangles to animated figures making basketball shots. The teacher aides became fascinated with Logo. Spontaneous groups formed around the aides working on very large Logo projects.

Calculator activities were drawn from the available books. We used inexpensive six-function calculators, beginning with order of operations and exercises to explore the square root and percent function keys. Calculator exercises were good anticipatory sets for computer problems and furnished analogies for discussions of computer processes.

The university campus provided a collection of experiences for science projects. The eclectic campus architecture presented a classification project. Experiments in air pressure, low temperature, and lasers were carried out in the science departments and technology center. The adjacent forestry college and journalism school were used for field trips on the environment and communications. A stream walk at a nature center in a nearby rural community was an outstanding science field trip. The students hiked the stream, observed the flora, collected specimens, and got very wet. Many different uses of computers, mathematics, and science were seen on a trip to the State University of New York Health Sciences Center. Students saw computers in the research laboratories, library, and operations center. They met the medical center staff and talked by a computer link with a student in South Africa. An important objective of the field trips was to show both career opportunities and computer applications.

Cultural Needs

Cultural awareness was incorporated in such a way that it was an extremely important aspect of the project. It has been suggested (Johnson 1990) that education and knowledge in general have to be driven by our culture. Therefore, we addressed the issue of culture in the educational experience. We used culture to drive and inspire the students to learn.

It is a truism that the environment shapes the mind, and vice versa. One's experience in American schools is an example of the interaction between school expectations and cultural environment. Culture is both horizontal and vertical. Horizontal in the sense that an individual must understand and be able to interact with one's environment, family, community, and workplace.

Vertical in the sense that an individual must understand one's origins and where they place one in the present.

In today's society, it still remains true that students with cultural and linguistic backgrounds that are different from those of the dominant culture do not yet have full access to quality education. Minority children are frequently placed in low-ability or remedial tracks, from which it is very difficult to move. Additionally, the course work is usually not demanding, precluding any opportunities to engage in higher-order thinking skills necessary to be successful in advanced-level mathematics courses in high school or college. It is possible to identify paradigms, theories, and knowledge claims that perpetuate inequality in mathematics achievement and that most teachers learn during their education (Johnson 1990).

John Ogbu (1987), an educational anthropologist, provides additional insight into how cultural differences affect on the educational enterprise. Ogbu states that the barriers for minority children's academic performance are threefold:

> First, whether the children come from a segment of society where people have traditionally experienced unequal opportunity to use their education or school credentials in a socially and economically meaningful and rewarding manner; second, whether or not the relationship between the minorities and the dominant group members who control the public schools has encouraged the minorities to perceive and define school learning as an instrument for replacing their cultural identity with the cultural identity of their "oppressors" without full reward or assimilation; and third, whether or not the relationship between the minorities and the schools generates the trust that encourages minorities to accept school rules and practice that enhance academic success. (Ogbu 1987, p. 334)

Unfortunately, most African Americans perceive that they are on the negative side of these three issues.

A career awareness component was also planned into the program to give the students insights into such professions as medicine, health technology, engineering, and environmental science and show the contributions of minorities in these areas. These careers were explored through the use of films, guest lectures, field trips, and workshops. As stated in the NCTM *Curriculum and Evaluation Standards,* "students' cultural backgrounds should be integrated into the learning experience" (1989, p. 68). Guests and university staff from a variety of backgrounds, such as engineers, artists, doctors, social workers, and television personalities, told why they chose their particular careers, how they got their education, and how they did their jobs, and they related the importance of mathematics and science to their work and careers. The series of presenters gave the students a view of people at different stages in their careers. Students asked good questions, and by the program's end, their questions showed greater insight—they even began to talk about their career choices.

Conceptual Framework

The conceptual framework that guided the project focused the learning and the teaching of mathematics and science and the use of technology on problem solving. It viewed the teacher as a problem poser and as a model problem solver. Any change in student learning was thus seen as resulting largely from this different approach in teacher behavior. The activities furnished the content in which these changes could take place. The goal was to enhance the teaching and learning of underrepresented students in mathematics and science by raising students' expectations and by selecting teachers who believed in the students and were willing to take the time to understand the students in the program. *The conceptual framework argued that there is a critical link among the activities of teachers, the learning of students, the curriculum, and the specific goal of the project—namely, the effective use of technology by teachers and students alike in problem solving.*

Achieving our goals, that is, resolving the educational contradiction discussed earlier, required that a series of actions be taken. These actions and the conceptual framework that guided them served as the basis of the solution for the problem as stated. In short, action involved—

- the effective use of existing software or the development of new software in mathematics, specifically at the middle school level of mathematics and science;
- the development of curriculum materials all geared for effective integration of technology in all phases of mathematics and science wherever possible;
- the involvement of parents and community to support the aims and activities of the project.

The actions were designed with certain outcomes in mind for both students and teachers. For the student the primary aim was to develop the ability to explore new, unfamiliar problems using appropriate strategies and skills. This required that students learn to analyze a problem, generate good conjectures, devise and carry out plans for verifying conjectures, and form the habit of making a critical review of results. Secondary aims were to develop computer literacy, programming skills in Logo, and associated mathematics and science concepts and skills.

For teachers to help students attain these goals, there are certain abilities these teachers must possess:

- A knowledge of the curriculum and its goal of problem solving
- Skills in recognizing valid problems worthy of being addressed
- An understanding of the process of problem solving and patience to experiment with ideas and strategies

- Skills in representing data graphically and developing useful and appropriate notation
- Organizational skills in managing the activities of a classroom in which computers are placed
- Skills in the use of computers and available software for problem solving
- Skills in communicating and sharing expertise with others.

Staffing the program with practicing teachers and undergraduate aides brought experience and enthusiasm to the project. A mathematically enriched program was developed around computer technology, action science, and related field trips. As Papert suggests, learning best occurs when children construct objects in an active environment. The most frequent comment from the student evaluations of the program was, "I enjoyed working on the computers; it was fun."

Evaluation Methodology

Syracuse University provides evaluation and research support for instructional development and other projects. The evaluation staff of the center designed an evaluation of the Mathematics and Science Microcomputer Project that provided information that—

1. documented the implementation of the program;
2. identified the program's strengths and suggested ways to improve the program;
3. determined if the program was viewed as a positive experience.

Subsequent evaluations will study the long-term impact of the program on whether students elect to study more mathematics and science in their school careers.

There were several different groups involved in the program: the seventh and eighth grade students from the Syracuse City School District, the parents of these participating students, and the Mathematics and Science Microcomputer Project faculty and administrative staff. Information from each group provided a somewhat different perspective on the program. Data collection procedures were developed to gather information efficiently from each group.

The participating students were surveyed regarding their perceptions of, and reactions to, the program and its components. The survey instrument included questions that required students to rate the science, mathematics/computer, and career awareness components of the program on the basis of (1) how informative they were, (2) how useful they were, and (3) how interesting they were. In addition, several open-ended questions sought information about what students perceived as the strengths of the program and what they would like to see changed.

The parents of the participating students were also surveyed. It was the intent of this evaluation to gather information on parental perceptions of the program as they were shaped by the impressions their children shared with them at home. As with the student survey, this instrument included a combination of rating and open-ended questions seeking to identify the strengths and weaknesses of the Mathematics and Science Microcomputer Project.

To gather the faculty's perceptions of the program, an open-ended questionnaire was used. The information served both to document the implementation of the Mathematics and Science Microcomputer Project and to identify what worked and what needed to be improved. In addition, the program's faculty evaluated each student on the following characteristics:

- Ability to participate in class discussions
- Ability to work in groups
- Attentiveness
- Attitude toward the instructor
- Attitude toward the subject
- Interest in the subject
- Prior preparation
- Verbal participation
- Willingness to try new things

Space was provided for open-ended comments regarding the faculty's perceptions of an individual student's strengths and weaknesses and recommendations to further the student's growth.

Evaluation Results

Survey forms were distributed to the students during the last week of the program. A total of 166 responses were received, a return rate of 77 percent, in approximately the same demographic proportions as represented in the program. Survey forms were also distributed to the parents during the last week of the program. A total of 136 responses were received, a return rate of 63 percent.

The participating students and their parents reacted very positively to the program. The evaluation findings suggested the following:

- Most students found the program's activities informative, useful, and interesting. They particularly enjoyed the career awareness sessions and the field experiences.
- Both students and parents felt that the program made substantial contributions to students' confidence in their academic preparation for the new school year.

- Students and parents would overwhelmingly recommend this program (98%) to their peers. As one student stated, "It's very fun [*sic*] and you can learn a lot of new and interesting things." In the words of a parent, "In general, it seemed a *wonderful* program, giving kids a chance to be associated with people who *like* mathematics and science. In public school, teachers don't have a background in those areas—consequently the kids think they hate it. This program is good because it allows the child to view the subjects in a positive way. This makes *all* the difference."

In general, the faculty, too, felt the program was successful. As one teacher responded, "I feel the program gave students additional information and enjoyment in science and mathematics. I enjoyed going into detailed activities on subjects the students chose. This type of detail can't be done in the normal school year and we covered topics that might not even be approached." Another faculty member suggested that the Mathematics and Science Microcomputer Project might increase students' interest in college preparatory programs.

The evaluation of students by the faculty suggests that overall the students were at least of average academic ability across all the demographics that were listed earlier. In addition, the faculty's identification of students' strengths and weaknesses and its recommendations for growth were shared with each student's academic-year teachers and school principal.

Summary and Conclusion

Mathematics is the most universal subject taught in school. Also, among the many subjects taught in schools, mathematics depends the least on the student's background and culture. Therefore, school mathematics should transcend cultural and economic diversity, and this is what we achieved in the Mathematics and Science Microcomputer Project. Minority students were given an opportunity to use the computer in a challenging way for something other than drill and practice, an experience they seldom receive but definitely need. The project's goals were to raise minority and economically disadvantaged students' expectations for success in mathematics and science with teachers who believed that success was possible. It is anticipated that once they are exposed to the microcomputer, students will be able to transfer these skills to other areas of academic work during the regular school year.

African American and economically disadvantaged people have always seen education as the best route to individual achievement. Today, an education, and a solid foundation in mathematics and science in particular, is far more necessary than ever before. In the early 1900s a person could become a successful member of society by stopping school and going to work as easily as by remaining in school until graduation. Today, that is not the case. The role of African American and economically disadvantaged students in mathematics, science, and technology is no longer an issue of equity; it is an issue of national survival.

The program had much success, especially in the eyes of the students and parents. Its major benefits included providing students with positive experiences in mathematics and science and increasing their interest in college preparatory programs. However, probably the most important message that African American and economically disadvantaged students received from this program was that careers where mathematics, science, and technology are involved are not alien to them or their world.

Bibliography

Becker, Joanne Rossi. "Differential Treatment of Females and Males in Mathematics Classes." *Journal for Research in Mathematics Education* 12 (January 1981): 40–53.

Blubaugh, William Leo. "Effect of a Semester of Instruction in Different Computer Programming Languages on Mathematical Problem Solving Skills of High School Students." *Dissertation Abstracts International* 45 (1984): pp. 30–86.

Carpenter, Thomas P., Mary Kay Corbitt, Henry S. Kepner, Mary Montgomery Lindquist, and Robert Reys. "Results of the Second NAEP Mathematics Assessment: Secondary School." *Mathematics Teacher* 73 (May 1980): pp. 329–38.

Johnson, Howard. "How Can the *Standards* Be Realized for All Students?" *School Science and Mathematics* 90 (October 1990): pp. 527–43.

Matthews, Westina. "Influences on the Learning and Participation of Minorities in Mathematics." *Journal for Research in Mathematics Education* 15 (March 1984): 84–95.

National Science Board, Commission on Precollege Education in Mathematics, Science and Technology. *Educating Americans for the 21st Century.* Washington, D.C.: National Science Foundation, 1983.

Nelson, Randle W. "Sex Differences in Mathematics Attitudes and Related Factors among Afro-American Students." Doctoral dissertation, University of Tennessee at Knoxville, 1978. (ERIC Document Reproduction Service, No. WE 005 296).

Ogbu, John. "Variability in Minority School Performance: A Problem in Search of an Explanation." *Anthropology and Education Quarterly* 18 (1987): pp. 312–34.

Papert, Seymour. *Mindstorms: Children, Computers, and Powerful Ideas.* New York: Basic Books, 1982.

Peng, Samuel S., W.B.Fetters, and Andrew J. Kolstad. *High School and Beyond: A National Longitudinal Study for the 1980's.* Washington, D.C.: National Center for Educational Statistics, 1981.

Wells, Gail Wilson. "The Relationship between the Processes Involved in Problem Solving and the Processes Involved in Computer Programming." *Dissertation Abstracts International* 42 (1981): SECA, 2009.

7

Blending Equity and Excellence: UMTYMP's Efforts to Reach All Students

Harvey B. Keynes

The University of Minnesota Talented Youth Mathematics Program (UMTYMP) is a statewide program aimed at providing an alternative educational experience for the most gifted among Minnesota's mathematically talented students. Typically identifying students in grades 6–8, the program supplies an intense academic environment and a culture of mathematics through a sequence of specially designed accelerated mathematics courses. These students attend a two-hour class one afternoon each week after school for thirty weeks from September through May. The students average five to ten hours of homework a week. In the first two years of the program (the high school component), students study Algebra I and Algebra II the first year *and* Geometry/Mathematical Analysis the second year. The high school component is taught by outstanding certified high school mathematics teachers and college-age teaching assistants. In addition, according to a 1984 Minnesota State law, two full years of high school mathematics credit is to be granted for completing Algebra I/II and Geometry/Mathematical Analysis (one year for each course). During the next three years (the college component), the students study calculus of one and several variables, linear algebra, and differential equations. The calculus component consists of special courses created for the UMTYMP students by the University of Minnesota at an accelerated honors level. UMTYMP calculus provides carefully monitored curriculum, syllabi, assignments, and examinations developed and taught by regular senior School of Mathematics faculty. Pursuant to the 1984 Minnesota state law, one full year (1/2 year per course) of high school mathematics credit

The author would like to thank Philip Carlson, Associate Director, Special Projects, and Laura Cavallo, Assistant Director, Special Projects, for their valuable suggestions in preparing this article.

is to be granted for completing of each of the two yearly calculus courses. More details can be found in Berger and Keynes (1990).

This paper will describe our recent approaches to dealing with the issues of equity and excellence and some of the activities by the program to enhance female participation. We first present an overview of UMTYMP's goals, design, and content and describe some of its unusual student-centered features. We will present some of the special issues facing female students as well as the program's philosophies and rationales behind the design of the intervention activities. We then discuss a pilot project that will help bring some newer technology-based instruction to UMTYMP and our initial attempts to use this technology-based course to attract more students of color to participate in UMTYMP. Next, we furnish a progress report on these efforts and outline our plans to maintain a long-term evaluation of these interventions and measure their effectiveness. We conclude with a description of related activities and some final remarks about the needs of mathematically talented students.

Goals, Design, and Content

In this section, we give a broad overview of the design and structure of UMTYMP. This description deals mainly with general attitudes and directions rather than specific details. The details can be found in Berger and Keynes (1990).

The overall goal of UMTYMP can be simply summarized. It is to furnish an intense, mathematically challenging, and dynamic academic program to mathematically talented and motivated students in a highly supportive and success-oriented environment. The curriculum is under constant discussion and revision to reflect the best and current ideas about all subjects, from algebra through calculus. Although classes are accelerated beyond the standard school schedule, the content, teaching styles, and support activities available to the students clearly enable *all* students to be highly successful. The program emphasizes work habits and homework as the key features to accomplishment rather than just classroom testing. An important aspect is to give students a sense of participating in a culture of mathematics and a sense of how this culture can help them in their careers and lives. Although the program's size requires some level of bureaucracy, the intellectual and curricular decisions are kept at the grass-roots level, namely, the highly qualified and highly motivated high school and university teaching staff.

Both curricula and textbooks in any subject are able to be revised and replaced in any year. The process involves a cooperative search and report by the teaching staff with the assistance of a program coordinator. As a result, textbooks and curricula in the high school courses tend to change about once every two years. The program encourages and supports "risk taking" at the curricular level when the intellectual goals will be better served. For example, the current geometry text, Morrow and Lang's *Geometry, a High School Course* (Springer-Verlag), reflects a modern and different approach to geometry that challenges both the students and the teaching staff. To support the teachers in this adoption, UMTYMP provides its own summer in-service opportunities and additionally compensates teachers serving on homework and examination teams.

We cannot overemphasize the importance of a broad and deep support system that has been developed by the program for the students and their families. These supports cover social and attitudinal aspects as well as academic issues and can be characterized as high-level nurturing to encourage all students to be motivated to succeed at their highest levels of capability.

One important aspect is the program's approach of making highly visible the "perks" from being interested in, and working hard at, achievement in mathematics. These advantages include the special admission procedures for UMTYMP students at several major universities and the striking success of UMTYMP students in science, mathematics, and engineering majors at prestigious colleges. Other aspects, to be described in detail below, include invitations to special popular lectures and events, a summer jobs program, and most recently, some innovative research experiences. Interest in the college achievements of UMTYMP graduates and the special opportunities for UMTYMP participants seem to be major factors in family support of the program.

To help align the goals of the program with the major goals of the NCTM (1989) *Curriculum and Evaluation Standards,* we provide the chart in figure 7.1. Under each standard is the aspect of the program that addresses these goals.

UMTYMP and the NCTM *Standards*

Standard 1:	MATHEMATICS AS PROBLEM SOLVING
UMTYMP:	• Curriculum emphasis on the process of solving problems. • Challenging problems are a part of all homework. • Applications of mathematics are emphasized.
Standard 2:	MATHEMATICS AS COMMUNICATION
UMTYMP:	• Provides exposure to a wide variety of mathematical concepts. • Mathematics notation is used in a manner consistent with higher education. • Students are required to write in good English.
Standard 3:	MATHEMATICS AS REASONING
UMTYMP:	• Geometry is approached through several proof patterns (synthetic, transformational, analytic). • Concepts and concept manipulation are expected in all courses, spiraling up to some proofs in college-level courses. • Conceptual problems are a part of all homework.

Standard 4:	MATHEMATICAL CONNECTIONS
UMTYMP:	• The total curriculum is intertwined with algebraic, geometric, and analytic approaches to mathematics. • Calculus is unified, and UMTYMP presents multivariable calculus using linear algebra, analysis, and geometry.
Standard 5:	CONTINUED STUDY OF ALGEBRAIC CONCEPTS
UMTYMP:	• Two years of algebra in the first year are followed by significant use of algebraic approaches in geometry and mathematical analysis. • Calculus and linear algebra continue to require algebraic approaches in the college part of UMTYMP. • Linear analysis relies heavily on linear algebra and algebraic concepts.
Standard 6:	CONTINUED STUDY OF FUNCTIONS
UMTYMP:	• Functions and the functional approach are a common thread running through all UMTYMP courses. Algebra is taught from an applications point of view leading to the formation of functions. The transformation view of geometry further develops the function concept. Mathematical analysis is approached as a study of classes of functions. Calculus is based on a more detailed analysis of functions, and functions are used in a variety of ways in applications. Linear algebra is based on the geometry and regularity of linear transformations.
Standard 7:	GEOMETRY FROM A SYNTHETIC PERSPECTIVE
UMTYMP:	• Although the second-year geometry component is the main place where these topics are covered, a geometric perspective is a common major thread in algebra, mathematical analysis, calculus, and linear algebra. In linear algebra, the entire approach rests heavily on geometric approaches in 2- and 3-space.
Standard 8:	GEOMETRY FROM AN ALGEBRAIC PERSPECTIVE
UMTYMP:	• Geometry from this algebraic perspective is studied in all UMTYMP courses. The geometry course includes this perspective. Linear algebra is strongly oriented to geometry from this perspective.
Standard 9:	TRIGONOMETRY
UMTYMP:	• In the algebra courses, trigonometric ideas are applied to triangles in a variety of problems. The function view of trigonometric functions is a major topic and approach in mathematical analysis. Trigonometric functions and ideas of periodicity continue to be used heavily in all the calculus courses.
Standard 10:	STATISTICS
UMTYMP:	• The construction of tables, charts, and graphs is taught in Algebra I. Measures of central tendency are also covered in algebra. The other areas of statistics are covered in Algebra II.
Standard 11:	PROBABILITY
UMTYMP:	• All the topics of probability proposed in the *Standards* are covered in Algebra I and Algebra II. Probabilistic problems and approaches are used in geometry, mathematical analysis, and all the calculus courses.

Fig. 7.1

UMTYMP's Approach to Equity and Excellence

UMTYMP strongly adheres to the principles and connections between equity and excellence as eloquently presented in *Everybody Counts* (National Research Council 1989). In particular, the program shaped its approach under a philosophy best expressed by the following principle: "Equity for all requires excellence for all; both thrive when expectations are high" (p. 29). The implementation of this principle in a program such as UMTYMP with its existing high standards and levels of achievement required careful planning and design. With initial support starting September 1988 from the Bush Foundation, the program designed several interventions addressing primarily informational, counseling, and support issues rather than the program's structure itself. They involved working with the families and the schools as well as with the students. The program believes that the need to encourage a supportive home and school environment cannot be overemphasized.

A statement on parental responsibility endorsed recently by both NCTM and MAA provides an excellent rationale for encouraging parental and school involvement in the intervention process. Distributed to UMTYMP parents, it asserts that "social, economic, or educational status of parents does not have as important an effect on their children's learning as what parents do with their children" (NCTM/MAA 1989, p. 6). We believe that family support is especially critical to UMTYMP. Recommendations to parents include discussing classroom activities with their children, furnishing a time and place for them to study, and participating in conferences if concerns arise. It also encourages parents to complement the program's efforts by becoming aware of the breadth of mathematics topics covered in their children's classes and becoming more involved in supporting the good work habits that are critical for success in UMTYMP.

The program decided that the best approach to equity and excellence would be to maintain the historical strength of the program and its curriculum and design a variety of support, social, and counseling activities that would enhance the learning environment and atmosphere for capable students who needed these extra interventions. We wanted these students to have the same benefits and opportunities that made the program so valuable to its traditional participants. Thus, we looked at many issues that concern school and social environments, ways to improve family support, study habits, and counseling needs and that in general provide a more sympathetic learning environment. We were convinced that the intellectual ability to deal with mathematics at the level required by UMTYMP was present in many students who were currently not in the program, including females and students of color. Following the models of the most successful intervention programs, we wanted the new atmosphere of the program to motivate and pique the interest of more students by having them discover the benefits and rewards—as well as the sacrifices—of the long-term culture of mathematics engendered by UMTYMP. We took the point of view that all these interventions should be carefully evaluated and that the program should try to identify which interventions worked and for what reasons. We also took care to note attitudinal shifts and anecdotal information that pointed to changes but that were more difficult to quantify. Because of limited resources, we planned to retain only those interventions that were most effective. Although these interventions were designed to influence female students, we decided to emphasize those that affect all UMTYMP students. We saw that an important secondary effect of our intervention was the improvement of the program for all students. Our results to date have clearly indicated that this approach can succeed quite admirably. We believe that the experiences developed in this type of project will help UMTYMP to initiate more successfully a similar project for students of color as described later in this paper.

Issues Facing Female Students

The lack of an appropriately supportive environment—in school and at home—as well as the lack of other girls in the class were important issues here. All too frequently one family indifferently requests the withdrawal of a girl performing near the top of her class while another tenaciously pleads to continue a boy who is struggling. Such cultural issues as socially coping with being a smart girl and failing to realize the impact of mathematics were other important aspects. Also, inconsistent and sometimes negative messages from the schools for participation in UMTYMP affected girls more severely. Finally, issues affecting personal esteem—lack of self-confidence in abilities, lack of involvement in even cooperative competitions—played a prominent role.

The issues associated with these factors are quite complex, but several features stand out. Students were primarily chosen to participate in the qualifying examination by the schools. It appears that the schools do not do as well at identifying mathematically talented female students as they do with male students. Even when a female student qualified, she was more likely to turn down admission than her male counterpart. Informal analysis indicated a lack of encouragement in both the schools and the home as an important factor. Once in the program, given equal ability and equal grades, females' persistence was lower than males'.

Bush Foundation Interventions Project

Social events have become a principal component of the intervention project and have been extended from girls-only in the algebra class to all levels of the program. Although the socialization began with female-only events such as bowling parties, movies, and pizza parties, the students have now expressed a strong desire for coed parties as well as a desire to become a part of the planning process. A letter from a concerned female geometry student suggested that rather than isolate the girls by sponsoring all-female parties, we should bring the girls into the mainstream with coeducational functions. She also suggested establishing a Student Board to plan the activities. Both of these suggestions are being implemented.

To support the perspective that UMTYMP not only provides a mathematics education but also encourages a culture of mathematics, older students participate in several social events that focus on college and career information. One such event was a calculus luncheon that featured a social hour followed by presentations from alumni on the impact of UMTYMP on their college experiences. The presenters included four females, two of whom are Ph.D. candidates, and three males. In an effort to provide women role models and possible mentors for female calculus students, a "shadow" program is being established whereby these young women will have the chance to experience women at work in various fields of study. A Learning Styles Workshop and Math Fun Fair round out the cultural experience of the younger UMTYMP students.

New testing procedures and special orientation meetings for girls who were successful in the qualifying examination were established. Activities to encourage and support girls who nearly qualified were also included in the intervention process. The near-qualifiers were invited to a variety of social activities with current UMTYMP girls. Including the near-qualifiers proved to be very successful in 1989–90, with a consistently large number of near-qualifiers attending the functions. Moreover, the pool of near-qualifiers in 1988–89 was an outstanding source of successful UMTYMP participants for 1989–90. These successes repeated for 1990–91.

On the basis of the literature and prior UMTYMP experience, it was decided that a class with a 50 percent enrollment of each sex would be most desirable. In 1988, three algebra classrooms were 50 percent female and two were all male. By fall 1990–91, all algebra and geometry analysis classes were 50 percent female.

In an effort to sustain parental involvement, UMTYMP has established a variety of academic year activities for parents and students, such as orientation sessions, a workshop on learning styles, and a mathematics fun fair. The academic progress of all students is monitored by a UMTYMP counselor, and continued emphasis is given to regular counseling contacts with the parents. The involvement of parents as a principal component of the UMTYMP support system has been one of the major successes of the various interventions. This aspect will be emphasized in the intervention program for students of color.

Curriculum Innovation: A New Technology-based Alternative Course

The program has always been aware that many highly capable students might benefit from variations on the standard UMTYMP curriculum. In fact, some of the successful students currently enrolled in UMTYMP found it helpful to repeat portions of a course already completed or possibly withdraw from the program for a portion of the year. The opportunity to enrich their background in a subject or enjoy a slower-paced period is frequently quite beneficial. Both before and after these periods, these students progress without difficulty through the standard UMTYMP courses. Sometimes they return to be top students after this period of change.

Another situation sometimes encountered is having capable and motivated students admitted to the program who subsequently discover some serious gaps in their mathematical skills or some major problems with their study habits. With intense counseling and major extra efforts, these students can progress in the program. But the extra work on top of the already heavy demands of the regular curriculum can be too burdensome, and sometimes these students are forced to withdraw.

To address both these situations, UMTYMP considered developing an alternative model that would maintain the same standards and basic curriculum of its current two-year high school program but provide a more enriched and more customized program over a three-year period. The somewhat slower-paced program would allow more time to enrich standard topics, furnish additional background on an individualized basis when necessary, and lessen somewhat the heavy homework commitment of the regular program. An additional tutorial class each week would encourage group work and appropriate study habits, supply focused activities on high-level problem-solving skills, and generally give whatever assistance and support is necessary. This model would ultimately allow students in the standard curriculum to transfer to this alternative if the need arose. Conversely, students completing the alternative curriculum should be able to handle the UMTYMP calculus program as well as the other students. Finally, it is hoped that the flexibility and individualization of the alternative program would encourage greater participation in UMTYMP by underrepresented groups. The opportunity to work individually with motivated and talented students and shape some of the skills necessary to succeed in UMTYMP would furnish another important dimension of equity.

The recent emergence of graphing calculators in the secondary school curriculum supplied another direction for this alternative model. With the current availability of the more user-friendly second-generation graphing calculators (e.g., the TI-81) and the general availability of better software, it was decided to emphasize technology in this alternative course. A new curriculum that will totally integrate graphing calculators will be developed. Computers will be used for classroom demonstrations, and the new materials will follow the guidelines suggested by the NCTM *Standards*.

Cray Foundation Intervention Project

Thanks to a commitment in March 1990 by the Cray Research Foundation to support a four-year pilot of this new alternative course, the program began to implement this project starting in 1990–91. One of the outstanding UMTYMP high school teachers, a Presidential Awardee nominee who is highly proficient in the use of technology in the classroom, agreed to develop the curriculum and teach the pilot course for the full three-year period.

With the background and knowledge from the Bush program, UMTYMP initiated a variety of activities during 1990–91 to ensure that the 1991–92 pilot class would achieve broad representation among students of color. Special contacts and commitments were sought from the schools, community organizations, and families. These include school visits, community visits (including churches and recreational groups), and parent meetings. A group of parents of students of color already in UMTYMP met with the program staff to provide information on how well the program is currently serving their students and how to structure new supports. Their strong belief in the value of UMTYMP and their active support for their children's involvement was striking. A larger advisory committee including community leaders and school contact persons was formed.

Most of 1990–91 was devoted to gaining credibility in the various communities and identifying a pool of potential students for the alternative pilot. Interviews of African-American students in UMTYMP by the major African-American newspapers were published. The Boys and Girls Clubs offered to have their tutors identify gifted children and offered the use of their clubs for testing and class purposes.

A special spring test was administered, and a residential summer enrichment institute was offered in summer 1991. The enrichment institute enabled the students to meet one another, become familiar with the instructor and the TI-81 graphing calculator, begin to understand how UMTYMP functions and its approaches and expectations about curriculum, and last but not least, study some interesting enrichment mathematics. A recently awarded NSF Young Scholars grant enabled the students attending the 1991–92 alternative course to meet monthly during the academic year for continued enrichment activities.

UMTYMP is very excited about this alternative course, just under way, and the potential benefits to its students and the program in general. We fully expect it to be yet another example of the enriching influences and improved learning environment that the equity and excellence activities have brought to the Talented Youth Program.

Progress Report

Although the Bush project is just completing its first three-year phase, initial results are very promising. In September 1988, partial implementation of the new testing procedures had already resulted in a larger and academically stronger applicant population. Even using a more difficult qualifying test and a higher cutoff from previous tests, the program admitted a class 20 percent larger than usual. The testing population was 44 percent female, and the entering class was 32 percent female. Despite higher standards, female scores in the algebra classes showed an overall improvement and were more evenly distributed than in the past. In the spring of 1989, an improved preregistration procedure was implemented that involved supplying schools with preregistration packets to be distributed to students. Parents then sent the preregistration form directly to the UMTYMP office. These new procedures created an equally strong applicant pool with better gender balance, and further gains in female enrollment were realized. In 1989, the testing population was 45 percent female and the entering class was over 40 percent female. That was an 83 percent increase over the 1987–88 academic year, the last year with no interventions. Furthermore, the females who preregistered in the spring scored as a group 2 points higher on the qualification test than the females who decided to register through the schools in the fall. The statistical difference between the means of the two groups is highly significant. These improved results persisted in 1990. Thus, these new procedures are identifying better-qualified female applicants. Table 7.1 summarizes results.

Table 7.1
Percent of UMTYMP Participants Who Are Female

	1986–87	1987–88	1988–89	1989–90	1990–91
Testing	43	42	44	45	46
Qualifying	23	29	30	43	42
Enrolled	21	22	32	40.3	40

Further examination of prior data suggested that the near-qualifier pool should be expanded to include those girls who scored within 10 points of qualifying for the program. (Previously, near-qualifiers needed to score within 4 points of qualifying.) This larger pool of the 1988 near-qualifiers either was given an opportunity to take a spring qualifying exam or received a letter urging them to retest in the fall of 1989. Of those forty-nine who retested, sixteen were accepted into the program. This is a remarkable qualifying rate of 32 percent, far higher than the historical qualifying rate of 8 to 10 percent. Similar results were obtained in following years. Overall, the new strategies to encourage near-qualifiers and implement more equitable enrollment appear to be quite successful.

Equal-enrollment classes have led to much improved and more supportive classroom dynamics for the female students. In fact, one class was clearly dominated by its strong and vocal female group in the first year. Together with the social opportunities and workshops, this structure has led to improved retention of girls.

For example, in the first year of intervention, the retention rate from Algebra I to Algebra II for boys and girls was essentially the same (91.9% for girls, 92.1% for boys). Prior to any interventions (the 1987–88 school year), the corresponding retention rates were 85.7 percent for girls and 90.7 percent for boys. The retention results for 1989–90 were especially impressive. In *every* subject in UMTYMP, there were fewer withdrawals by female students than by male students. Table 7.2 gives details.

Another measure of effectiveness is grade distribution. One major concern in a large intervention program is that the new students will not be as academically able as the previous classes. The program was convinced that the additional female students would maintain and enhance the academic structure of UMTYMP. Table 7.3 indicates that with the current environment, female students are achieving at a level equal to or higher than their male counterparts.

The success of these interventions can sometimes be measured in a different context by observing the behavior of the students. At the end of each spring term, the algebra students are given the opportunity to request classmates for their next year's geometry class. In forming these classes, we were able to honor almost every request. One class chose to stay together into geometry. On the first day of fall classes, returning students excitedly greet friends and rush to the bulletin board where the class lists are posted to find out with whom they will share their geometry class. Students frequently call friends who have decided not to return to the program to encourage them to

Table 7.2
Retention Rates for UMTYMP Students

Retention Rates	Bush Group I	Bush Group II
Algebra I to Algebra II	Fall '88–Spring '89	Fall '89–Spring '90
Males	92.7%	92.4%
Females	93.4%	98.1%
Algebra II to Geometry	Spring '89–Fall '89	Spring '90–Fall '90
Males	79.2%	76%
Females	70.6%	80%
Geometry to Math Analysis	Fall '89–Spring '90	
Males	98.3%	
Females	100%	
Math Analysis to Calculus	Spring '90–Fall '90	
Males	64%	
Females	32%	

Table 7.3
UMTYMP Grade Distributions

Bush Group I	Algebra I Fall 1988	Algebra II Spring 1989	Geometry Fall 1989	Math Analysis Spring 1990
# of students	M = 81 F = 38	M = 76 F = 37	M = 59 F = 28	M = 58 F = 28
% obtaining A	M = 50.6 F = 39.5	M = 44.7 F = 35.1	M = 54.3 F = 60.7	M = 48.3 F = 53.6
% obtaining B	M = 39.5 F = 55.2	M = 40.8 F = 51.3	M = 39 F = 35.8	M = 38 F = 25.1
% obtaining C	M = 3.7 F = 2.6	M = 6.5 F = 2.7	M = 5.1 F = 3.6	M = 5.1 F = 3.6
% obtaining D	M = 6.2 F = 2.6	M = 7.9 F = 8.1	M = 1.7 F = 0.0	M = 8.6 F = 17.9
Bush Group II	**Algebra I Fall 1989**	**Algebra II Spring 1990**		
# of students	M = 79	F = 53		
% obtaining A	M = 41.8 F = 54.7	M = 39.7 F = 40.4		
% obtaining B	M = 44.3 F = 39.7	M = 45.3 F = 46.1		
% obtaining C	M = 6.4 F = 3.8	M = 1.4 F = 5.7		
% obtaining W (Withdrawal)	M = 7.6 F = 1.9	M = 13.7 F = 7.7		

participate. Friendships created in UMTYMP often extend outside the university classroom even though students live miles apart.

The overall aim of the Bush Intervention Project is to create a supportive environment in UMTYMP that encourages capable females to involve themselves seriously in mathematics. Its success is indicated by the following quotes:

- "I realize the importance of the early encouragement I got in UMTYMP; I will pass on that encouragement to other girls as a role model and through volunteer projects." (*A UMTYMP graduate and former UMTYMP Teaching Assistant*)
- "Last year was our daughter's first and I saw the benefits in light of her growth and development and love of mathematics." (*A UMTYMP parent*)
- "She has never thought of herself as 'very good' at mathematics, having been in a class with other equally good students. However, the reality of it is now dawning on her and it has really made a difference to her self-confidence." (*A UMTYMP parent*)

Other Aspects of UMTYMP

Based on a pilot program in the summer of1988 and with partial support from the NSF Young Scholars Project, a two-week special summer institute for current UMTYMP students and a new summer enrichment institute for prospective UMTYMP girls were offered in 1989. Both institutes, held on the university campus, were special enrichment-oriented programs that teach interesting topics different from those in the usual curriculum while incorporating problem-solving techniques and mathematics-league competition skills. The residential summer institute was designed for continuing UMTYMP students, whereas the nonresidential enrichment institute was a commuter program designed to increase the interest of female near-qualifiers and spring-qualifying students to participate in UMTYMP. Along with enrichment activities, the institutes provided industrial tours, career information, and socialization activities for students. Outstanding high school teachers from the academic-year UMTYMP classes as well as University of Minnesota mathematicians were involved in the program.

Two special institute activities took place in 1989. Students who had completed at least one year of calculus were given an opportunity to work with John Hubbard, a mathematician with special interests in computer applications to geometry, on his innovative educational software. Several world-class mathematicians associated with the Geometry and Visualizations Project and UMTYMP cohosted an extraordinary closing luncheon event for both institutes, local high school teachers, and other potential students. This luncheon was preceded by a highly informative talk on the geometric aspects of tilings by William Thurston, of Princeton University, and followed by an exceptional presentation (with videos and films) by the founder of fractals, Benoit Mandelbrot. His talk on the "Fractal Cosmos" was so alluring that many students asked for his autograph at the conclusion!

The third summer institute and second summer enrichment institute were held in June 1990. Following the same format as in 1989, the teachers in each of the two summer programs developed problem-solving skills through a variety of mathematical challenges. A very dynamic mathematics-league competition was carried out for the two weeks. One teacher and a college-level teaching assistant worked with the girls in the enrichment institute on projects that had a strong geometric flavor. Both groups shared some speakers on mathematics topics. Tours presented many examples of mathematics and science applications, and women scientists and technicians gave several talks and demonstrations. In addition, the enrichment institute girls had a tour of the university's Department of Chemical Engineering led by women majors and heard an intriguing talk by a physics professor, who demonstrated the connections of mathematics to physics by a number of experiments.

In 1986, UMTYMP initiated a summer Minnesota job-placement program for UMTYMP graduates currently attending colleges and universities. It has become a permanent feature of the program. In the summer of 1990, twenty-nine résumés were received, resulting in several placements at UNISYS, EWA Company, and the University of Minnesota Computer Science Department. Six other students (including two freshmen) were hired by the University of Minnesota Geometry and Visualizations Project, with two students returning for their second summer. These six students were involved in a variety of undergraduate research programs and were described as "keen superstudents . . . full of wonder and questions about computers and mathematics." Their performance was outstanding, and they have been asked to participate again in the expanded summer 1991 research program. Several students are now regarded as junior staff by the Geometry Project.

The statistical data from the program over its fifteen-year history and evaluations from UMTYMP graduates indicated that UMTYMP was overall doing an excellent job in educating its students. The postsecondary record of UMTYMP students is striking. College opportunities for both admissions and scholarships are impressive. Several major universities, including the University of Chicago, Harvard, and MIT, give preferential admission consideration on the basis of the prior successes of UMTYMP graduates. Female graduates fare especially well. Most choose difficult engineering and science majors at leading schools and easily persist in their programs. Approximately 84 percent (36 of 43) of the female students are entering some field of mathematics, science, or engineering. Although nationally only 2.2 percent of incoming freshman intend to major in one of the physical sciences or computer science, 51 percent (22 of 43) of UMTYMP females and 61 percent (113 of 184) of the males choose one of these majors. In alumni surveys, they testify to the importance of UMTYMP in their intellectual growth and in developing self-confidence. The program appears to patch the severe "pipeline leakage" too often seen for female students at the undergraduate level. See table 7.4 for more details.

Some Final Remarks

An interesting aspect of running a program for talented students is the recurrent question from many different sources about the need for such programs. Our society commonly beleives that there are many good programs in the schools for such students and that these students will survive and be successful no matter what curricular opportunities are available. Much of UMTYMP's information and experiences would strongly challenge both of these assumptions. Many UMTYMP parents and students have indicated that there were very few quality opportunities available in their schools and that the students were frequently extremely bored, lacking motivation or underachieving. A recent report (*Minneapolis Tribune,* 3/23/91) quoting David

Table 7.4
Survey of Majors of UMTYMP Graduates

Males (184)			Females (43)		
Major(s)	No.	%	Major(s)	No.	%
Mathematics	44	24	Biology	8	19
Physics	39	21	Mathematics	8	19
Electrical engineering	24	13	Physics/astrophysics	6	14
Computer science	19	10	Biochemistry	4	9
Engineering	10	5	Chemistry	3	7
Chemistry	9	5	Chemical engineering	2	5
Aero engineering	8	4	Computer science	2	5
Biology/bio med	7	4	Economics	2	5
Mechanical engineering	6	3	Engineering	2	5
Chemical engineering	5	3	Undecided	4	10
Economics	5	3	Majors occurring once	9	21
Philosophy	4	2	TOTAL	50 *	
English	3	2			
Political science	3	2			
Business	2	1			
Biochemistry	2	1			
Education	2	1			
Premedical	2	1			
Undecided	21	11			
Majors occurring once	16	9			
TOTAL	231 *				

*Some alumni have multiple majors.

MAJORS OCCURRING ONCE
Males: Actuarial Science, Air Traffic Control, Civil Engineering, Communications, Comparative Religion, Geology, History, Russian, Science, Sociology, Finance, East Asian Studies, American Literature, French Literature, Film Studies, Natural Science

Females: Civil Engineering, Clinical Psychology, Community Health Education, English, History, Mechanical Engineering, Natural Science, Japanese Studies, Italian

Sadker and Myra Sadker, among others, disturbingly pointed out the continuing differential treatment of boys and girls in school. It was particularly noted that "bright girls were the most neglected group of students because they tended to be quiet and undemanding." The need for supportive environments for mathematically talented female students remains a major problem in Minnesota at the current time.

UMTYMP has developed in its fifteen years of operation into one of the most comprehensive and sophisticated long-term educational programs for mathematically talented students in the United States and even throughout the world. Its goals have been to provide a totally mathematically oriented learning community in a supportive and positive environment. It has attempted to create a culture of mathematics for its students and thus has dealt with issues of educational philosophy, leadership, and environment. It has learned that to address the educational needs of these talented students appropriately, it must also deal with their families, their schools, their communities, and their lives.

One important issue that UMTYMP has learned to address is the psychological and emotional aspects of being talented. The program has worked hard to give its female students a (totally justified) sense of self-confidence and an honest sense of accomplishment and self-worth. These girls are aware of their abilities to succeed in demanding academic situations involving both boys and girls. The program consistently emphasizes that with significant efforts on their part, their achievements are typically among the best of any students. This message may be an important factor in the exceptional success of female UMTYMP graduates in science, mathematics, and engineering at some of the most demanding colleges and universities. A strong sense of self-direction seems to persist in many aspects of their lives. UMTYMP will continue to survey this important issue over time to see if the effect can be more substantively documented.

The program's experience with the Bush Intervention Project has resulted in many beneficial aspects. Its activities have provided a solid foundation for our new intervention program for students of color. In addition, these experiences have led to interventions and support activities that have improved the learning environment for all UMTYMP students. Finally, the program has become acutely aware of the need to remain innovative, flexible, and responsive to creating new procedures and structures as we learn more about educating such extraordinary students. For UMTYMP, equity has strengthened excellence.

References

Berger, Thomas, and Harvey Keynes. "The Challenge of Educating Mathematically Talented Students: UMTYMP." AMS/CBMS Issues in Mathematics Education 1 (1990): 11–32.

National Council of Teachers of Mathematics. *Curriculum and Evaluation Standards for School Mathematics.* Reston, Va.: The Council, 1989.

National Council of Teachers of Mathematics/Mathematical Association of America. "Parent Involvement Essential for Success in Mathematics." *Focus* [MAA Newsletter] 9 (June 1989): 6.

National Research Council. *Everybody Counts: A Report to the Nation.* Washington, D.C.: National Academy Press, 1989.

8

Algebra Transition Project: A Work in Progress

Sue Stetzer

Imagine an inner-city algebra 1 class in which students work cooperatively to solve problems. Journal writing is an integral part of the course. Each student has a scientific calculator. There is a mobile cart with a computer and a projection device at the front of the room. Algebra is taught in the context of real-world problems and applications. The teacher acts as facilitator and coach. Expectations are high and success is in the air.

Is this a fantasy? No, it is an algebra 1 class in Philadelphia's Algebra Transition Project. The Algebra Transition Project is jointly funded by the Philadelphia Schools Collaborative and the School District of Philadelphia. Its goal is to increase the number of students taking—and succeeding in—algebra 1. To date, it appears to be working.

Background

The School District of Philadelphia is an urban district of almost 200 000 students; 77 percent of the total enrollment, K–12, is African American, Asian, or Hispanic. In 1989, there were 9 340 high school students taking algebra 1 and 5 224 high school students taking general mathematics. Approximately 38 percent of the algebra 1 students and over 52 percent of the general mathematics students did not pass their mathematics course.

The high schools in the school district fall into two categories: magnet and comprehensive. The magnet schools are racially desegregated and have special programs that attract students from throughout the city. The comprehensive high schools are neighborhood schools that take students from the immediate school neighborhood; they may be racially segregated. Many of the twenty-one comprehensive schools suffer from the woes that befall dysfunctional inner-city schools. At one comprehensive high school, for example, only 26 percent of all ninth-grade students received mathematics credit in 1989.

That same year, average daily attendance at another comprehensive high school was just under 60 percent. In response to the problems of the comprehensive high schools, the Pew Charitable Trust funded the Philadelphia Schools Collaborative in April 1989 to renew, restructure, and revitalize the comprehensive high schools.

In 1989, the National Council of Teachers of Mathematics released its *Curriculum and Evaluation Standards for School Mathematics* (NCTM 1989). The *Standards* calls for a core curriculum for all high school students that includes topics in algebra, geometry, data analysis, and trigonometry. This is in sharp contrast to current practice, which tracks the most able students to academic courses and denies equitable access to quality mathematics to the less able. If everyone is to be exposed to meaningful mathematics, then general mathematics is no longer an appropriate high school mathematics course. More students should be enrolling in—and successfully completing—algebra 1 because "first-year algebra is the keystone subject in all of secondary mathematics" (Usiskin 1987, p. 428). And according to *Everybody Counts* (National Research Council 1989, p. 7):

> More than any other subject, mathematics filters students out of programs leading to scientific and professional careers. From high school through graduate school, the half-life of students in the mathematics pipeline is about one year; on average, we lose half the students from mathematics each year, although various requirements hold some students in class temporarily for an extra term or a year. Mathematics is the worst curricular villain in driving students to failure in school. When mathematics acts as a filter, it not only filters students out of careers, but frequently out of school itself.
>
> Low expectations and limited opportunity to learn have helped drive dropout rates among Blacks and Hispanics *much* higher—unacceptably high for a society committed to equality of opportunity. It is vitally important for society that *all* citizens benefit equally from high-quality mathematics education.

It was in this context that a small group of mathematics department heads from comprehensive high schools assembled for a series of meetings in the office of the Philadelphia Schools Collaborative in the fall of 1989. The agenda was to design a program to improve student access to quality mathematics.

A vision of the Algebra Transition Project emerged:

- Focus on the transition summer between the eighth and ninth grades.
- Offer students a six-week summer school experience whose entire thrust is mathematics.
- Provide a program rich in prealgebra experiences, taught in nontraditional ways.
- Follow this program with a course in algebra 1 reflective of the recommendations of the *Curriculum and Evaluation Standards.*

This program should make it possible for students of below-average ability to pass algebra 1 at a significantly higher rate than the citywide average.

With this vision in mind, comprehensive high schools applied to be part of the project. The application required specific commitments. Each school had to identify four interested teachers. At least two of the teachers had to be high school teachers and at least one teacher had to be from the local middle school. The teachers had to agree to participate in staff development, to change their instructional strategies to implement the program, and to teach in the summer program as well as during the school year. Principals and schedulers had to indicate their willingness to support this project, particularly with respect to assigning successful students to specific algebra 1 sections taught by participating teachers. This project also had to be coordinated with restructuring efforts sponsored by the Philadelphia Schools Collaborative. Six schools submitted proposals; three were selected.

Components of the Program

The Algebra Transition Project has five components: curriculum materials, staff development, student recruitment, the summer program, and the school-year component.

Curriculum Materials

The Algebra Transition Project teachers selected the University of Chicago School Mathematics Project (UCSMP) *Transition Mathematics* textbook as the basis for the summer curriculum (Usiskin et al. 1990). This book focuses on comprehension, relying on the scientific calculator for tedious computation. It has many examples of nonroutine problems, with heavy emphasis on real-world applications. Students must think about and understand what they are doing. For example, the concept of variable is introduced through the idea of patterns (Usiskin et al. 1990, p. 146):

1 person has $2 \cdot 1$ eyes.

2 people have $2 \cdot 2$ eyes in all.

3 people have $2 \cdot 3$ eyes in all.

4 people have $2 \cdot 4$ eyes in all. . . .

Let p be any natural number.

p people have $2 \cdot p$ eyes in all.

Subsequently, students are asked to apply their understanding of patterns by giving a pattern for the following instances (Usiskin et al. 1990, p. 184):

> If the weight is 5 ounces, the postage is 5¢ + 20 · 5¢.If the weight is 3 ounces, the postage is 5 + 20 · 3¢. If the weight is 1 ounce, the postage is 5¢ + 20 · 1¢.

Transition Mathematics devotes five of its thirteen chapters to the meaning of operations, foregoing the traditional emphasis on mechanical drills. An entire chapter is devoted to "Patterns Leading to Addition" wherein the operation is explained using the "putting-together model for addition" and the "slide model for addition." Within the chapter, the uses of addition appear as motivation to develop both skills and properties. These ideas are finally applied in the solution of simple equations that involve addition. Students using the book must learn how to read a mathematics text to become independent learners. In short, the book provides a solid foundation for algebra.

The teachers identified specific topics in *Transition Mathematics* to cover during the summer. They chose chapters 4, 5, 7, and 8—sections dealing with integers, variables, and linear functions. They felt that students familiar with these concepts would be prepared to study algebra 1. These topics were supplemented by enrichment units on using the scientific calculator, data analysis, computer applications, and geometry.

The teachers chose UCSMP *Algebra* (McConnell et al. 1990) for the school year algebra 1 textbook. It incorporates all the features valued in *Transition Mathematics.* In addition, its strong geometry component lays the foundation for future success in geometry.

The Philadelphia Schools Collaborative purchased the necessary textbooks. In addition, the collaborative bought Texas Instrument's TI-30 Challenger calculators for each student and mobile computer systems—computer and printer, cart, and projection viewer—for each classroom. Each teacher received an overhead projector calculator and the software that accompanies the UCSMP textbooks.

Staff Development

The new materials required changes in teaching behaviors. The teachers knew that they could not succeed with their students by using traditional, teacher-centered instruction. A program of twenty-five hours of paid staff development gave teachers training in new instructional techniques. The topics were identified by the teachers themselves and were intended to provide alternatives to lecturing.

Cooperative learning was the major focus. Throughout the *Curriculum and Evaluation Standards* are recommendations to use cooperative learning. Many teachers wanted to know what it was, to feel comfortable using the method, and to incorporate it into their classes. An introductory session on cooperative learning was held, followed by an all-day workshop in which teachers worked in groups, modeling cooperative learning as they learned specific techniques for its implementation.

The staff development also included sessions on using technology, both calculators and computers. From the first chapter, the UCSMP materials rely on scientific calculators. Such total integration was new for all teachers, and training was required. Teachers also needed training to enable them to take advantage of the power of graphing software available for computers. The other staff development topics included the following:

- Using the UCSMP materials
- What is algebra in light of the *Curriculum and Evaluation Standards?*
- Using manipulatives
- Reading and mathematics
- Critical thinking

The staff development was conducted during two full Saturday training sessions plus six after-school sessions of 2.5 hours each.

To sustain changes in teaching behaviors, the project required teams of teachers from within a school. The teams provide peer support and the opportunity for ongoing dialogue. This enables teachers to reflect on new instructional strategies as they practice them. During the planning phase, the teams met regularly to plan for the summer program. Throughout the summer, the teams met for an hour each day to debrief, discuss successes and failures, revise lesson plans, and discuss students' progress. Because all the teachers were covering the same material at essentially the same pace, these meetings were rich sharing sessions. During the school year, the teams meet monthly in users' groups.

Student Recruitment

The target population was exiting eighth-grade students who would not normally either take algebra 1 as ninth-grade students or be expected to pass algebra 1. The targeted students were in the 16th–49th percentiles in mathematics computation.

Students completing the summer program would take algebra 1 with the same teachers they had had in the summer. Each cluster of four teachers could accommodate up to 90 students and still keep class size down in the summer. The combined enrollment for the three high schools was targeted at 270 students. Expecting some attrition, each high school planned for two full sections of algebra 1 for September. Because maximum class size is contractually set at 33 students, the goal was successful completion by 66 students in each high school.

Each high school recruited from its pool of entering ninth-grade students. Every effort was made to present this opportunity as a privilege, open to a select group of handpicked students. The students were offered a rich summer experience in prealgebra and two additional inducements: each successful student received a $50 United States savings bond and a high school elective credit. Elective credit, rather than mathematics credit, was given so as not to reduce the number of additional mathematics courses the students would need to graduate from high school. Free tokens for transportation were supplied for all students.

Each middle school distributed flyers announcing recruitment meetings for parents. The meetings gave parents and children an opportunity to meet the high school staff and visit the schools. Often this contact doesn't occur until the first day of high school, if at all, and certainly is never done with the relaxed informality that is the luxury of a summer project. The teachers actively courted parents and invited them to visit classes, chaperone trips, and become part of the project.

The target was 270 students; 262 were initially recruited. Unfortunately, on the opening day of the program only 170 students appeared. A local jobs program recruited a number of potential students; other students were required to attend remedial summer school or chose employment. However, of those who began the program, most stayed to complete it. Average daily attendance was 153, and 155 savings bonds were issued to successful students.

The Summer Program

The summer staff consisted of three clusters of four teachers each; a coordinator, who is the mathematics department head at one of the participating high schools; and a disciplinarian/counselor. Twelve high school seniors, four at each high school, worked as paid mentors for the summer. These mentors were strong mathematics students who assisted teachers, tutored the students, and served as role models.

Community College of Philadelphia (CCP) was selected as the summer site. It was a neutral facility that provided twelve classrooms and a conference room with a telephone that could be used as the program's office—all on one floor. Auditoriums, a computer lab, and a cafeteria were also available. The relationship between CCP and the Algebra Transition Project was more than that of landlord and tenant. Students had an initial orientation and tour as well as opportunities to interact with CCP students, who served as positive role models. A member of the mathematics department at CCP worked with the program. He was a mathematics consultant when the staff had questions, and he also conducted an informal SAT mathematics preparation class for some of the mentors.

Community College of Philadelphia is closed on Fridays during the summer, so the Friday schedule was a series of off-site educational and recreational trips. The three educational trips were to the Franklin Institute Science Museum, where students saw the interconnections between mathematics and science; to the Philadelphia Art Museum for a look at geometry in art; and to the Free Library, where all students obtained a library card and viewed the mathematics videotape *Cosmic Zoom.*

The daily schedule for students, Monday through Thursday, appears below:

9:00– 9:10 Homeroom
9:10–10:10 First period
10:10–10:25 Break

10:25–11:25 Second period
11:25–11:35 Journal writing
11:35–12:15 Group problem solving and homework

The homeroom teacher was also the first-period teacher. First period consisted of one hour of instruction from *Transition Mathematics.* The teachers selected the sequence of topics to be covered, and all teachers followed the agreed-on sequence and schedule.

After the break, the students rotated through the team of four teachers from their cluster for a second hour of instruction, meeting each instructor once a week. Each teacher presented enrichment modules, which were coordinated within the team. These open-ended explorations included computer activities, data analysis activities, games, and projects. One teacher used M&M's with her class to teach the concepts of mean, mode, and median. Another teacher in the team then taught the students how to use a spreadsheet to organize and graph the data they had gathered. Students also used the data gathered from their small bags of M&M's to predict the color distribution of M&M's in a one-pound bag. Letters were written to the Mars Company asking about the intended color distribution, and comparisons were made between the data gathered in class and the "ideal" bag of M&M's.

Following the enrichment modules, ten minutes were set aside for daily journal writing by both teachers and students. The journals were intended to help students think and communicate mathematically. Some days, the students were given specific prompts for their journals, such as "Ben was absent today. Write him a note telling him what he missed." Other days, the journals were open-ended. The journals also served to furnish insight for the teachers into the students' lives. The following excerpt is from a student's journal entry on the final day of the program:

> I am writing this to say, I think that this project is a very good idea. I learned a lot of neat things, such as the "op-op property" and "instances" and other things that I think will help me in high school, and further on in life. I have other reasons why I like this program. I liked the program, also because of the teachers. To me this made my summer more interesting.

The student day concluded with a homework–problem-solving session lasting forty minutes. This provided a time frame in which to attempt a variation of Uri Treisman's work at Berkeley (Treisman 1990). Treisman, on sabatical leave from the Unviersity of California, Berkeley, was in the Philadelphia area at Swathmore College during the 1989–90 school year. He had spoken to Philadelphia teachers about his work with minority students in the Professional Development Program at Berkeley. The Algebra Transition Project teachers took Treisman's work as a model for replication. Students used the time to work in small groups on homework problems and enrichment activities. Teachers and student mentors were available to assist the groups.

The emphasis was on building student support groups, which would work cooperatively on challenging problems. This period was intended to be rigorously enriching rather than remedial in focus, and the hope was that the groups would continue throughout the school year. After the student day ended, the teachers had an additional hour for planning, peer support, administrative tasks, and curriculum development. This time furnished an opportunity for teachers to share their experiences and support each other.

To maximize continuity in September, the students were assigned to one of their summer teachers for algebra 1. As a result, both teachers and students saw the summer as a beginning rather than as a casual experience unrelated to the serious work of high school. The summer furnished an opportunity for students to bond with their teachers as well as with their classmates before entering high school. Once ninth grade began, these connections eased the students' transition to high school.

The middle school teachers served several functions in the program. They were vital to the recruitment efforts because they, rather than the high school teachers, knew the potential students. Their presence provided the students with familiar faces at the outset. They shared their expertise in teaching early adolescents with their high school colleagues. Most significantly, the dialogue between middle school and high school teachers has improved articulation between the schools.

School Year Component

When school began in September 1990, 146 students were enrolled in algebra 1 classes taught by the same teachers they had had in the summer. The remaining students either had transferred to other schools or had insurmountable scheduling conflicts. The teachers began the school year by issuing the UCSMP *Algebra I* textbook on the first day. There were no start-up delays because the students already knew their teachers and their classmates. The students were also familiar with the format of the textbook, the style of the teacher, and the high expectations that the teacher set for the class. Classes were organized in cooperative groups, a model begun in the summer. The algebra classes are continuing to implement the UCSMP program begun in the summer. Algebra is taught in the context of real-world applications. Each student has a calculator that is used for classwork, homework, and testing. Homework is assigned daily; there are both a nightly reading assignment and a series of topical questions. Computer demonstrations and journal writing are integrated as time permits. The following student journal entry demonstrates the impact of the Algebra Transition Project:

> The Algebra class is different from my other classes because I knew my teacher before the school year and we're really good friends.

In the middle schools, the teachers are teaching *Transition Mathematics* to their eighth-grade classes, using the textbooks purchased for the summer program. The hope is that the middle school teachers will produce students who are better prepared for future success in algebra 1, ultimately obviating the need for the summer component.

Teachers within each of the three clusters meet regularly to discuss their successes and failures with the new materials. There are also monthly "users' group meetings" that bring together the entire summer staff to discuss the curriculum and students' progress. Teachers share their experiences and ask questions about specific sections of the text. The discussions vary from the general—"My students are doing much better now that we're in chapter six. They seem to understand what I'm talking about"—to the specific—"I think the section on the distance formula, with absolute value inside the square root symbol, was particularly hard to teach."

Preliminary Evaluation Results

During the summer, parallel 44-item pretests and posttests were given using instruments taken from the supporting materials that accompany *Transition Mathematics.* The average increase from pretest to posttest scores was 11.7 items. Those students with the lowest pretest scores showed the most gain, indicating that perhaps the Algebra Transition Project best served those who needed it most.

At this writing, it is too early to know whether these students will, in fact, pass algebra 1 at a rate that exceeds the city average. However, the interim data are very encouraging. At the end of the first quarter, 95 percent of these students attended mathematics class at least 90 percent of the time and 85.6 percent passed algebra 1. These figures are significantly higher than comparable data for ninth graders in general; the impact of the figures is all the more dramatic given the low percentile rankings of the target students.

The students took a nationally normed, citywide algebra 1 midterm examination, and 97 percent of them passed. The algebra 1 teachers were fearful that the students' daily reliance on calculators throughout the course might prove a handicap on an examination where calculators were not permitted. This did not prove to be a problem.

Ongoing monitoring will determine students' success in algebra 1. Additional data will be collected to evaluate students' attendance, promotion rates, and success in future mathematics courses. Those students who worked as mentors during the summer and were tutored for the mathematics component of the SAT have also benefited, showing an average increase of 67 points on their mathematics SATs; their verbal scores increased an average of 17 points. (*Note*: These students were tutored for the SAT, not because they were deficient, but because they hoped to raise their scores and be more competitive.)

Observations and Issues

The Algebra Transition Project presented a group of teachers with the rare opportunity to create a vision and then implement that vision. Atypically for budget-pressed school systems, resources were provided as necessary to pay for staff development, textbooks, computers and calculators, student trips and incentives, and so on. The teachers had the vision to shape a summer program, the power to define a specific body of content, and the freedom to take risks and experiment in the summer with new instructional techniques. The teachers have confronted the very real problem of students entering high school with little hope of anything but the general mathematics track. They have defined a way to better serve these students and, in the process, have empowered themselves by placing themselves at the center of solving the problem. Their sense of ownership of the program and their investment in their students has certainly sustained an attitude of high expectations.

The students have had a smooth transition to high school and are doing well in algebra 1. In two of the high schools, there were not enough summer students to fully populate two algebra 1 classes. Small groups of students who were not part of the summer program were added to the classes. Although they had a noticeably difficult start, over time they have had the summer students as their mentors and are no longer distinguishable from the summer students. The classes have sustained the message of the summer: students will work and will succeed.

The parents see the summer teachers as advocates for their children. In situations where their children were having difficulty with a teacher in another subject, the parents have come to the algebra teachers for help and mediation. The Algebra Transition Project parents have been meeting throughout the year to maintain a group identity.

The teachers who worked in the summer program are very proud of their achievements:

> Morale remained high and conversations continued to reflect energy, optimism, even excitement. Some were surprised by the students' efforts to work in their classrooms. Many were finding this experience to be far more enjoyable/rewarding than their school years. Friendships developed, as did humor. Flexibility enabled the team to respond capably to the unexpected. . . . It was particularly satisfying to see those teachers who initially questioned their inner resources gain confidence and enjoy successes. (Zubrow 1990, p. 62)

Although the program has met with success, not everything is perfect. The benefit of hour-long classes, like those held during the summer, is sorely missed. Forty-five-minute periods seem inadequate and abrupt; teachers find it difficult to regularly find time for journal writing and computer demonstrations within a lesson. At least one school is considering double periods for algebra 1 next year as a possible solution.

The initial vision of the project included tutoring, cultural activities, and ongoing student mentoring throughout the school year. These activities never came about. The students would benefit from a meaningful program of support beyond algebra class. Such support would also strengthen the social network created by the project. A possible prototype for such a high school program occurs at Treisman's Professional Development Program (PDP) in Berkeley, which involves "the students' working on problem sets whose mathematical content is similar to the course material being covered but whose questions require thoughtful answers rather than rote responses" (Stanley 1991, p. 147). The teachers are beginning to discuss how to adapt and implement the model locally.

The question of what comes next naturally arises. The bonds between teachers and students are so strong that the teachers are reluctant to send their students on to traditional geometry or second-year-algebra classes. The project was never envisioned to be longitudinal, however, and the students are destined for more traditional courses in future years. The hope is that other teachers in each of the high schools will move toward the recommendations of the *Curriculum and Evaluation Standards.* Such a move would be prompted in part by the demands of the students who have had a quality experience in algebra 1 and in part by the model of success provided by the Algebra Transition Project teachers.

Although the summer program will be repeated with a new group of students during the summer of 1991, two factors may contribute to significant changes. The first is that some of the students eligible for the summer will have already used *Transition Mathematics* as eighth-grade students. It may be necessary to rethink the summer content and move beyond *Transition Mathematics.* The second is that it appears possible to connect the program to an afternoon jobs program that would provide employment for eligible students. That connection should help attain the target number of 270 students.

The Philadelphia Schools Collaborative has received calls from parents, teachers, and principals from schools not participating in the project. They want to know how they can get involved in the project. Positive interest in the study of mathematics can be infectious. Children and teachers who would normally gravitate to general mathematics have caught the excitement and want to be part of the Algebra Transition Project.

Conclusions

The students who participated in the Algebra Transition Project were students who were originally destined for general mathematics courses in high school. A select few might have found their way into a "slow" algebra 1 course that would have taken them two years to complete. The Algebra Transition

Project provided these students with a rich and challenging mathematics program, some additional supports described above, and the expectation that they could and would succeed. Armed with newly found confidence and mathematics skills, these students are succeeding in algebra 1 instead of turning off in general mathematics. These are not exceptionally talented students. Their mathematics scores on nationally normed tests are significantly below the 50th percentile. If these students can succeed in a rigorous algebra 1 course, it is incumbent on teachers and school systems across the country to examine instructional and tracking practices and restructure these practices as necessary to make algebra 1 accessible to all. The message is clear. All children can be successful in algebra 1.

References

McConnell, John W., Susan Brown, Susan Eddins, Margaret Hackworth, Leroy Sachs, Ernest Woodward, James Flanders, Daniel Hirschhorn, Cathy Hynes, Lydia Polonsky, and Zalman Usiskin. *Algebra.* University of Chicago School Mathematics Project. Glenview, Ill.: Scott, Foresman & Co., 1990.

National Council of Teachers of Mathematics. *Curriculum and Evaluation Standards for School Mathematics.* Reston, Va.: The Council, 1989.

National Research Council. *Everybody Counts: A Report to the Nation on the Future of Mathematics Education.* Washington, D.C.: National Academy Press, 1989.

Stanley, Dick. "PDP High School Mathematics Workshop Model." *Mathematics Teacher* 84 (February 1991): 146–147.

Treisman, Philip Uri. A Study of the Mathematics Performance of Black Students at the University of California, Berkeley. In *Mathematicians and Education Reform,* edited by Naomi Fisher, Harvey Keynes, and Philip Wagreich, pp. 31–47. Chicago: American Mathematical Society, 1990.

Usiskin, Zalman. Why Elementary Algebra Can, Should, and Must Be an Eighth-Grade Course for Average Students. *Mathematics Teacher* 81 (September 1987): pp. 428–38.

Usiskin, Zalman, James Flanders, Cathy Hynes, Lydia Polonsky, Susan Porter, and Steven Viktora. *Transition Mathematics.* University of Chicago School Mathematics Project. Glenview, Ill.: Scott, Foresman & Co., 1990.

Zubrow, Judith. "The Algebra Transition Project: July 2–August 9, 1990." Unpublished documentation report. Philadelphia: Philadelphia Schools Collaborative, 1990.

Part 3
Changing How Teachers Teach

The improvement of students' learning of mathematics involves changes in students' behavior as well as in teachers' behavior. The NCTM's *Standards* implies that teachers need to change from traditional instructional practice. The changing of teachers' instructional behaviors encompasses formal training, experiences with new modes of teaching, sensitivity to students' learning needs, and new curriculum structures.

The chapters in this section reflect the foregoing factors. Camerlengo's paper provides one approach for a teacher education program. Bobango's, Heid and Jump's, and Thompson and Jakucyn's chapters describe ways in which curriculum can be structured to provide more effective instruction to students. In addition, these authors raise the issue of teachers' having to be more sensitive to their students' needs in mathematics learning.

Santiago and Spanos furnish in their chapter an underlying theme to the section: the development of communication skills in the mathematics classroom. They include the features of language that teachers need to be aware of, as well as strategies for improving classroom discourse.

9

Mathematics Specialist-Teacher Program: An Intervention Strategy for All

Vivian M. Camerlengo

The face of the United States is changing. In 1988, African Americans represented 12 percent of the U.S. population and 2 percent of all employed mathematicians and scientists. Hispanics, the fastest growing minority group, represented 9 percent of our population and, again, only 2 percent of scientists and mathematicians. Asian Americans, who made up 2 percent of the U.S. population, represented 6 percent of the mathematical and scientific workforce (Task Force 1987). By the year 2000, one out of three of our citizens will be nonwhite (Steen 1987). By extension, we could say that at least one out of three of our mathematicians and scientists also should be nonwhite.

Although the loss of U.S. students from mathematics, science, and technology has been widespread, it is especially noticeable among minority groups. In 1986, Hispanic and African American students represented a total of 6 percent, 3 percent each, of all the students enrolled in graduate programs of mathematics and science. Foreign students, many newly arrived or first-generation Asian Americans, accounted for 22 percent of those enrolled in advanced study of mathematics and science; 65 percent of those in advanced programs where knowledge of mathematics was essential were white graduate students (approximately 51% male and 14% female).

In an attempt to investigate possible sources of the unbalanced representation, researchers have related several variables to mathematicss achievement. Participation in mathematics courses, parental influence, sex of the child, and levels of cognitive development have all been examined. Among the complexities and interactions of variables studied, many explanations and much confusion exist. There is clear agreement on one point, however, and that is that disperity begins to appear early in students' elementary school careers. A former secretary of education, Lauro F. Cavazos (1989), reported that minority groups in general and

African American and Hispanic chilsren in particular "tune out" (before they drop out of) mathematics as early as the fourth grade.

Mathematics Education Perspective

If mathematics continues to be a white, middle-class, male dominated domain, we will not have the technological and scientific work force that is necessary to survive and compete in the twenty-first century, nor will the average citizen, whose daily life will demand mathematical competence, be adequately prepared. International comparisons reveal that U.S. students do not score as well as most of the students from nations with whom we compete (McKnight et al. 1987). The *Science Report Card* (Educational Testing Service 1988) reports that only half of the United States' seventeen-year-olds have any "sophisticated understanding of mathematics" and that a majority of thirteen-year-olds are "poorly equipped for informed citizenship and productive performance in the workplace, let alone postsecondary studies in science." Fortunately, many efforts, including federally funded intervention programs; the recent establishment of *Curriculum and Evaluation Standards* for the K–12 mathematics curriculum (NCTM 1989); a redefinition of "basic skills"; the integration of mathematics, science, and technology; and an examination of the preparation and qualifications of the mathematics teacher, are under way across the country to correct this trend.

As we enter the last decade of the twentieth century, we embrace a changing world that demands a reconceptualization of the role of education in general and mathematics education in particular. There is a growing need for a kind of mathematics education that will be sufficient for the numeracy skills of this general citizen as well as be appropriate for the demands that the work force of the future will face. All children must be steeped in this kind of enriched mathematics education from the earliest elementary grades. This article describes an intervention program designed to provide children this opportunity through the use of specialist teachers of mathematics in the upper elementary school grades.

Rationale

The Mathematics Specialist-Teacher Program (MSTP), a response to this national call, was implemented on local levels with district teachers, through federal and private funding, during the calendar years 1987–1990.

The MSTP represents an attempt to minimize the attrition of elementary school children in mathematics by increasing the knowledge and awareness of, and the fondness and enthusiasm for, mathematics on the part of their teacher of mathematics. A particular focus of the program is problem solving, which has become a focus of general discussion among educators. We want all our citizens to be able to demonstrate problem-solving skills when faced with unfamiliar situations. These skills are not easily developed, so they should be part of children's mathematical education from the earliest years. The mathematical notion of model building permits children to begin with problems in their immediate environment that are meaningful to them, interact with components that reflect the problems' constraints, and, eventually, manipulate mathematical symbols that define the particular task.

Many teachers of mathematics, especially in the elementary school where shopkeeper arithmetic has become entrenched, are themselves not fully aware of the current fields of mathematics that have developed since the turn of the century nor of the curriculum that the mathematics education community is recommending. The results of a 1986 national survey of mathematics and science education revealed that 74 percent of the teachers who were then teaching mathematics in our country had never taken a course in mathematical problem solving or the history of mathematics (Weiss 1987). On a more local level, data collected from another group of teachers who participated in an earlier study showed that 82 percent of them had not taken a course in number theory, probability and statistics, or the history of mathematics.

The power and beauty of mathematics reside in its parsimony, ubiquity, connectedness, and applicability. Children need to be exposed to these characteristics of mathematics at an early age if they are to increase their wonder and continue their participation in the study of mathematics, and they must be taught in a way different from the way most of their teachers were taught mathematics. Rote memorization of algorithmic processes is not what elementary school mathematics is about these days—nor will be in the next century!

Assumptions of MSTP

The Mathematics Specialist-Teacher Program operated on the following assumptions:

1. Mathematical competence and problem-solving ability are attainable, enjoyable, and exciting for *all* children.
2. Problem-solving ability in mathematics is directly influenced by experience as well as by affective and cognitive factors.
3. Children can retain and enhance their natural curiosity about mathematics in the world around them, increase their self-confidence and motivation, and develop an accurate sense of reasoning when the mathematical talent, interest, and awareness of the elementary school teacher of mathematics are maximized;
4. Mathematics at the elementary school level should be taught *only* by those teachers who have available current knowledge as well as the desire and interest to teach mathematics—retrained, revitalized elementary mathematics teacher-specialists.

On the basis of these assumptions, MSTP included instructional strategies that focused on—

- problem solving;
- the use of visual imagery, symmetry, and geometric transformations to develop concepts;
- the involvement and use of a child's kinesthetic sense to deepen and extend concepts;
- the connection and integration of the unifying themes and components of the mathematics curriculum.

Program Goals and Components

The MSTP project gave teachers experiences in problem solving and in building, applying, and interpreting mathematical and scientific models that are appropriate for implementation with their students in the upper elementary school grades. The preparation of MSTP teachers took place over a two-year period and involved six graduate courses for which the participants earned eighteen graduate credits. Courses required in the program included the following:

Problem Solving and the Nature of Mathematics

Algebra Extended

Geometry Revisited

Probability and Statistics in the Elementary School

Enrichment through Diagnosis

Integrating Technology/Culminating Seminar

The MSTP staff realized that teachers must, in general, conform to the objectives set by their individual districts. Accordingly, their preparation was accomplished in conjunction with their own mathematics curriculum guide and with frequent consultations with the district supervisor of mathematics.

Goals

The major goal of the MSTP program was to develop competent mathematics specialists who could serve as knowledgeable, enthusiastic role models in selected elementary schools in Washington, D.C., and Pennsylvania. A secondary objective was to have the specialists research, develop, and implement problem-solving lessons, in conjunction with appropriate heuristic strategies, that could be integrated with their curriculum guide. A third objective was to establish a network of mathematics specialists that would comprise support groups who could continue to interact after the project's funding had ceased.

An additional goal of MSTP was to determine the anxiety level of the teachers and their mathematical ability to solve problems at the beginning and end of their two-year experience with the program. This last objective also included an appropriate measure of mathematical anxiety that was administered to their students during the two-year period. Also, standardized mathematics achievement scores (CTBS) of students of the specialists were collected for analysis.

Population

Although there were two populations of teachers involved, thirty-four from Washington, D.C., and fifteen from Pennsylvania, it was primarily a District of Columbia project. This was due largely to the funding and administrative arrangements. MSTP was funded for three years by a federal grant for Washington, D.C., teachers; additional funds were obtained from the Exxon Foundation for one year. These were used, in conjunction with the federal funds, to investigate the feasibility of extending the specialist network and implementing the model with a group of teachers of Hispanic children in Pennsylvania.

In the District of Columbia, teachers themselves first determined if they would be interested in becoming mathematics specialists; informational flyers were distributed to the elementary schools inviting those who wanted to know more about the program to a Saturday meeting. From the teachers who were still interested, a committee (comprising a District of Columbia mathematics supervisor, principals, and the MSTP project director) selected two from each school. Teachers in Pennsylvania were selected by the regional office. The determination to have two teachers from each school, wherever possible, was made so that they could work as a team, function if one was absent, collaborate, and share experiences.

Since the District of Columbia was going to generate a position in the elementary schools for a "certified mathematics specialist," one of the criteria used in the selection process was the probability that the individual would stay with the program for the duration. All but two teachers in Cycle I did; these two finished their missing courses with Cycle II teachers.

Cycles

Although funding was provided for three years, progression through the sequence of six courses took only two years. Thus, the program was designed so that Cycle I teachers began in the fall of 1988 by enrolling in one of the six courses; they took another in the spring and two courses during the summer of 1989, when they were joined by Cycle II teachers. Cycle I and Cycle II teachers, therefore, were enrolled in four courses together.

The intention here was to begin building an intimacy among the teachers that would produce a self-sustaining support group. Indeed, some of the most remarkable experiences during those courses came from cooperative groups that capitalized on individual strengths. MSTP also used Cycle I teachers during the training of the second group so that a peer, albeit mentor, relationship would be established and encouraged to continue after the program ended.

Curriculum

The curriculum was designed according to the suggestions in the working draft of the *Curriculum and Evaluation Standards for School Mathematics* (NCTM 1987) in conjunction with the needs of the teachers and according to a needs-assessment statement by the district. A problem-solving project, funded by Title II monies in 1986, was designed to address the mathematical needs of teachers (Camerlengo 1987). At the conclusion of the one-year program, 20 percent of the teachers in the project subsequently enrolled in a calculus 1 class and received an A or B grade; 10 percent elected to take other mathematics courses; and 12 percent enrolled in a second workshop on problem solving.

The needs-assessment document also recommended that seminars and studies be conducted that address teacher attitudes and rapport with students. Studies on the interaction of student and teacher attitudes have indicated that teachers have a more favorable attitude toward mathematics as a process than students do but that students have more favorable attitudes toward the role of mathematics in society than their teachers do.

To this end, attitudes of the teachers were measured by the Mathematics Anxiety Rating Scale before and after their MSTP project experiences. The applicability and utility of mathematics were woven throughout the six courses with materials and films from the Challenge of the Unknown program (AAAS 1986).

The course in problem solving was designed to be taken first, since all mathematics in the subsequent courses would be presented in that light. Polya's model of problem solving and heuristic strategies, as well as information processing models (Newell and Simon 1972; Greeno 1973) along with Kantowski's assumptions (1981), Charles and Lester's lesson plans for problem solving, and activities from Krulik and Rudnick (1988), Greenes (1987), and Gardner (1988), were investigated.

Consider the following examples from the course. The first defines an exercise for the student who knows the "*P*lease *m*y *d*ear *A*unt *S*ally" mnemonic for "parentheses, multiplication, division, addition, and subtraction." The second and third illustrations include problems that involve the individual in constructing multiple solutions even when the order of operations is known.

1. Evaluate: a. $8 - 4^2 =$ ______ and $(8 - 4)^2 =$ ______.
   b. $7 + (3 \times 5) =$ ______.
2. Using the symbols +, −, x, and ÷ and parentheses, obtain a true statement for each of the following (AAAS 1986):
   a. 4 3 6 = 13
   b. 4 3 6 = 1
   c. 12 3 1 = 6
3. Using all the symbols of operations above, plus the square root symbol, write expressions for the first five natural numbers by using only 4s in *as many ways as you can.*
   a. 1 = ______
   b. 2 = ______
   c. 3 = ______
   d. 4 = ______
   e. 5 = ______

One principle that guided the program was an emphasis on connections within mathematics—connections among the apparently (and superficially) disparate components of the discipline. Consider the situation when teacher-participants first encountered this problem:

Suppose there are several people at a party and each one shakes hands with each of the others. If twenty-eight shakes are exchanged how many people were at the party?

First reactions included these:

"I know there's a formula for it, but I forgot it."

"I've seen this problem before—I think—but it wasn't quite like this."

"If we knew the number of people and wanted to get the number of handshakes, we could just list all the combinations, but not the other way around."

"I don't even know where to start, except maybe with a smaller number of shakes."

With some coaching from the MSTP staff, this last suggestion was encouraged and the "acting out" strategy was adopted. It was quickly seen that the solution structure could be represented as $7 + 6 + 5 + 4 + 3 + 2 + 1$, and the following generalization was offered by one of the teachers:

"It's easy—either you do the Gaussian thing where you fold the series in half, add the first and the last term, the second term, and the term next to the last, and so on . . . and you end up with . . . let's see . . . 3 and 1/2 groups of 8, or you simply start with one less than the number of people $(n - 1)$ (because you can't shake hands with yourself) and continue adding $(n - 2)$ and so on, down to 1."

Everyone seemed quite delighted with the discovery until a challenge was issued:

"Yes, OK, but what do you do if it is a very large number of handshakes; do you have to add all those numbers?"
"There's some formula for that; what is it?"

"You don't need to memorize formulae; you've seen something like this before," was the response from the MSTP staff member. At this point, the instructor went to the board and wrote (without talking) the results of an earlier investigation that involved the sequence of triangular numbers 1, 3, 6, 10, 15, . . . and a general represention of its *n*th entry, (1/2) $N(N + 1)$.

N (Ordinal Location)	n (Triangular Number Itself)	N' (Number of People)	n' (Number of Handshakes)
1st	1	2	1
2d	3	3	3
3d	6	4	6
4th	10	5	10
5th	15	6	15

"Is there any relationship between the number of handshakes and the ordinal position of the terms in the triangular number sequence?" the instructor queried. Amidst the expressions of wonderment, the instructor interjected, "Now, put that on hold for the time being and recall what you learned about permutations and combinations. At that time you discovered the number of combinations of eight objects if you only consider two at a time." The teachers' remarks indicated various levels of discovery and understanding.

"WHAT is that? Do you mean triangular numbers are related to Pascal's triangle and combinations?"
"I guess that makes sense. . . . If you have eight people at a party and only two shaking hands at a time, youd have a particular instance of the 8C2 model!"
"Yes, and we related those to the number of sides and diagonals in a convex polygon."
"What?. . . I don't remember."

At this point the MSTP facilitator went to the board and elicited the responses from the class needed to complete the following chart:

s (Number of Sides of Polygon)	d (Number of Diagonals)	$s + d$ (Sum: Number of Sides + Number of Diagonals)
3	0	3
4	2	6
5	5	10
6	?	15
7	?	?
8	?	?

The teachers were amazed! It was a little more difficult and even more astonishing for them to realize (but they eventually did) that the mathematical principles involved in the handshake problem are the same as those required to compute a part of the probability expresssion for a Punnett square in Mendelian genetics or to solve another, seemingly unrelated problem.

If you wanted to ensure a winning ticket at jai alai or the greyhound races on a quinella bet, how many tickets would you have to buy?

Of course, purchasing the required number of tickets to ensure a win does not mean that you necessarily finish with more money than you started. A discussion concerning mathematical expectation and probability was begun. Thus, the mathematics teacher-specialists, in training, began to see that Polya's first heuristic, "understand the problem," means more than understanding the words; it means seeing the underlying relationships and mathematical structure inherent in apparently distinct situations. It also demands a consideration of tangential as well as common intersecting issues.

Some interesting results occurred when the teachers first began to implement these ideas in their classroom. One teacher reported what happenend when the children were given the following problem. Pandemonium reigned!

Place the first six natural numbers in the circles in the diagram in such a way that each side totals to the same sum.

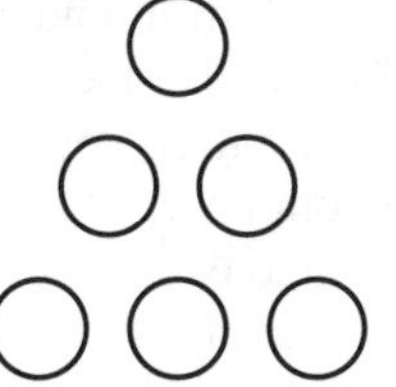

Students' first reactions included the following remarks:

"You have to tell us the sum."
"It can't be done."
"You never showed us how to do this. How are we supposed to know?"

After sustaining the cognitive dissonance for a few minutes, they got to work. Now the comments were as follows:

"I got it, it's 11."
"No, it's not. Our sum is 9."
"Well, this is really crazy. We have 10!"

The noise continued until the teacher began to organize her students' results in a pattern. She went to the projector, used different-colored pens, wrote the numbers the students gave her, and arranged them in an organized format.

1	1	2	4
6 5	6 4	5 3	3 2
2 4 3	3 2 5	4 1 6	5 1 6
sum = 9	10	11	12

This observation led the students to discuss the role of the vertices in the arrangement and to conjecture about entries other than the first six natural numbers. Some of the students even tried fractions!

Other students began to question how you could obtain an odd sum by either adding three odd numbers or adding two even numbers and one odd number. The teacher responded with, "Oh, you mean what happens when you have $3(2n + 1)$ or when you are adding $2(2n) + (2n + 1)$." Of course, the students were now into an algebra lesson.

Another inspired sixth-grade teacher began a unit on the golden mean with a measuring activity. She had the children measure, in metric units, their "navel height" and total height. Young adolescent children enjoy noticing and talking about their bodies, so this was a hit at the beginning. When the data were arranged in tabular form, the children inspected all the entries, made conjectures about values that seemed "out of line," and began a discussion on the notions of measures of central tendency and measures of dispersion. The teacher asked the children, "Besides the individual numbers, which vary depending on the person, what do you notice about the entries in the height, navel height, and ratio columns?" After a few not-too-insightful responses, one student said, "The first two columns stand for centimeters, the ratio column does not." Wonderment in the class grew when the students realized that the entries in the last column all clustered about the same number, regardless of the height of the individual.

Name	Height (*H*)	Navel Ht. (*N*)	*H/N*
Sasha	152	94	1.62
Garen	163	100	1.63
Robbie	156	99	1.58
Olinda	141	87	1.62
Alleni	150	93	1.61
Patricia	147	88	1.67
Brendan	161	98	1.64
Marcel	162	100	1.62

Seizing the opportunity, the teacher then discussed the fascination that surrounds this particular number and the number's many manifestations in the natural world, including a pinecone, a pineapple, a sunflower, and a nautilus shell. Thus, she integrated science with real-world applications under a mathematical generalization.

In addition, the Fibonnaci sequence 1, 1, 2, 3, 5, 8, . . . was revisited and, with the aid of a calculator, ratios of the $(n + 1)$th term to the nth were calculated and arranged in a table,

nth term	$(n + 1)$th term	$(n + 1)/n$
1	1	1
1	2	2
2	3	1.5
3	5	1.666 . . .
5	8	1.6
8	13	?
13	21	?

Two weeks before this lesson the teacher had the children act out the "bunny problem": children began in a kneeling down posture to represent a "just born" bunny, moved to sitting in a chair when they were "one month old," and stood up when they were old enough to "have their own bunnies." When their attention was focused on the table of Fibonacci ratio data, students could see a relationship at the conclusion of the exploration. Conjectures were elicited by the teacher concerning what number the ratio, $(n +1)/n$, seemed to "approach"! Almost all children responded with something like, "Oh!—it looks like that funny number, 1.6 . . . something, that we saw before!" This was pretty sophisticated mathematics for sixth-grade children who had not been tracked in the "college prep" strand!

Suggestions for Replication

For MSTP implementation to progress, there must be MSTP personnel and administrative involvement. In Washington D.C., MSTP staff *and* the mathematics coordinator visited the principals to discuss their students' mathematics performance on standardized tests, ascertain administrators' and students' needs, and explain the program. Involving administrators in the planning stage seems to be a crucial ingredient in a successful implementation process. Time must be budgeted during the planning phase of the project to establish social and professional relationships with administrators, and several visits should be made to each school to get a "feel" for school climate. Other specialist programs throughout the nation report similar results when working with school districts.

Also, parental involvement is needed. The fact that mathematics in the elementary school today is not what it was demands a reorientation on the part of the parents. Family Math sessions are being conducted by specialist programs in other regions of the nation, and initial reports indicate that they are very sucessful.

Through prior projects and programs, the MSTP staff had built a relationship with administrators and teachers. Principals had enrolled in previous workshops and were committed and excited about implementing a program where "mathematics" and "thinking" were not antonyms. The state supervisor of mathematics and the project director had collaborated on other projects.

Summary

Overall, the MSTP program appears to have been successful in achieving its goals. Thirty-four teachers are now, or soon will be, operating as mathematics specialists in twelve schools in Washington, D.C. Teachers anxiety levels have decreased, their knowledge and awareness of mathematics have increased, and preliminary analyses suggest an interesting result on the part of their students. The most recent standardized test results indicate that the childrens' problem-solving scores have increased.

One of the Cycle I teachers, whose sixth-grade class was videotaped by the National Academy of Science during a mathematics lesson, was invited to serve on the Mathematics Science Education Board. In addition, she maintains her teaching duties and informally "directs" the sharing sessions when the specialists meet.

Additional funds are being sought to continue preparing more teachers for elementary schools that were not represented during the first two cycles. This step should expand the local network and contribute to the sense of community support that is necessary to maintain a high degree of enthusiasm for mathematics teaching. It also should ensure that our knowledgeable, reflective, well-prepared mathematics specialist-teachers will provide a safe and charted course for *all* our students.

References

American Association For The Advancement of Science. *Challenge Of The Unknown.* Washington, D.C.: The Association, 1986.

Camerlengo, Vivian. *Title III: Final Report.* Washington, D.C.: D.C. Public Schools, 1987.

Cavazos, Lauro. "Building Bridges for At-Risk Children." *Principal* 68 (May 1989).

Educational Testing Service. *The Science Report Card.* Princeton, N.J.: ETS, 1988.

Gardner, Howard. "Creativity: An Interdisciplinary Perspective." *Creativity Research Journal* 1 (December 1988).

Greenes, Carole, and George Immerzeel. "Problem Solving: Tips for Teachers." *Arithmetic Teacher* 34 (January 1987): 26–27.

Greeno, James. *Creative Processes in Representations of Problems.* Report GK-4 tothe Office of Naval Research. Washington, D.C.: Office of Naval Research, 1987.

———. "The Structure of Memory and the Process of Solving Problems." In *Contemporary Issues in Cognitive Psychology,* edited by R. L. Salsa. Washington, D.C.: V. H. Winston & Sons, 1973.

Kantowski, Mary Grace. "Problem Solving." In *Mathematics Education Research: Implications for the 80's,* edited by Elizabeth Fennema, pp. 111–26. Alexandria, Va.: Association for Supervision and Curriculum Development, 1981

Krulik, Stephen, and Jesse Rudnick. *Problem Solving: A Handbook for Elementary School Teachers.* Newton, Mass.: Allyn & Bacon. 1988.

McKnight, Curtis C., F. Joe Crosswhite, John A. Dossey, Edward Kifer, Jane O. Swafford, Kenneth J. Travers, and Thomas J. Cooney. *The Underachieving Curriculum: Assessing U.S. School Mathematics from an International Perspective.* Champaign, Ill.: Stipes Publishing Co., 1987.

National Council of Teachers of Mathematics. *Curriculum and Evaluation Standards for School Mathematics.* Working Draft. Reston, Va.: The Council, 1987.

Newell, Allen, and Herbert A. Simon. *Human Problem Solving.* Englewood Cliffs, N.J.: Prentice Hall, 1972.

Steen, Lynn A. "Mathematics Education: A Predictor of Scientific Competitiveness." *Science* 237 (July 1987): 251–52, 302.

Task Force on Women, Minorities, and the Handicapped in Science and Technology. *Changing America: The New Face of Science and Engineering.* Washington, D.C.: National Science Foundation, 1989.

Weiss, Iris. "How Well Prepared Are Science and Mathematics Teachers?" Paper presented at the annual meeting of the American Educational Research Association, Washington, D.C., April 1987.

10

Meeting the NCTM Communication Standards for All Students

Felicita Santiago
George Spanos

Consider the following prealgebra classrooms, one taught in the traditional "chalk and talk" manner, the other taught with an eye toward increasing opportunities for students to communicate mathematically.

Class A: The teacher begins with a review of the homework assignment on the commutative property (if a and b are integers, then $a + b = b + a$). The students are asked to read their answers to each problem. Next, the teacher, through a combination of writing on the board ("chalk") and lecturing ("talk"), gives several examples, using such terms as *commutativity, variable,* and *integer,* and then asks for questions. The students are interested in knowing the solutions to the problems not included in the answer section of the textbook, and the questions on the next test. After answering a few questions, the teacher assigns homework, consisting of more textbook exercises, and dismisses the class.

Class B: Another teacher, presenting the same property, begins by having the students review their homework in small groups and report their results to the class. The students are encouraged to ask questions of each other concerning their application and understanding of the property; to explain how they arrived at their answers; and to state the property as coherently as possible. The teacher then introduces the property through the use of materials that invite the students to work in pairs, doing exercises in which they try to identify paraphrases of the property or cowriting sentences using such terms as *commutivity, variable,* and *integer.* Following the paired reports to the class, the teacher demonstrates the property with manipulatives, such as blocks or Cuisenaire rods, which represent the terms of the equation and can model commutivity. The students then model the teacher's demonstration. For homework, they are asked to write in their journals what they learned that day.

The teacher of class B recognizes that students learn mathematics by talking, writing, and reasoning about mathematics. They are able to develop proficiency in mathematics language through active use of that language in meaningful contexts. Ideally, this teacher has learned the approach through interaction and team teaching with language teachers to help both mainstream and limited English proficient (LEP) students in the mathematics class.

The Need for Communicative Mathematics Teaching

Mathematics and language educators have begun to implement what may be termed a *communicative approach* to the teaching of mathematics. Two general concerns provide the impetus for this approach. First, mathematics educators have begun to recognize the need for all students to become mathematically literate. The recent curriculum, evaluation, and teaching standards of the NCTM (1989 and 1991) recommend increased attention to communication in mathematics programs.

Second, mathematics educators and language educators (see, for example, Cocking and Mestre [1988]; Cuevas [1984]; Dale and Cuevas [1987]; Secada and Carey [1990]; and Spanos, Rhodes, Dale, and Crandall [1988]) have become concerned with the special linguistic and cultural needs of minority-language students. Such students are faced with the task of learning or relearning mathematics before they have developed competence in academic English or familiarity with the mainstream classroom context.

Learning to communicate mathematically is one of NCTM's (1989, p. 8) five major standards for instruction and evaluation in grades K–12. Whereas the authors argue that all students can benefit from listening, speaking, reading, writing, and demonstration activities (pp. 26–28, 78–80, 140–42), they add the following statement about nonnative speakers of English:

> Students whose primary language is not the language of instruction have unique needs. Specially designed activities and teaching strategies (developed and implemented with the assistance of language specialists) should be incorporated into the high school mathematics program so that all students have the opportunity to develop their mathematical potential regardless of a lack of proficiency in the language of instruction. (p. 142)

More recently, NCTM (1991) has emphasized the need for tools that enhance classroom discourse (p. 34), has recommended a range of discursive activities for teachers and students (pp. 3554), and has stressed the need for networking between language arts and mathematics educators (p. 193). This stress on communication and information sharing has created a partnership role for language and mathematics educators in developing materials, teaching methods, and assessment procedures that promote the integration of language learning with mathematics learning.

Communicative Teaching at the International High School

One program featured in a communicative mathematics videotape developed by the Center for Applied Linguistics (1990) is at the International High School at LaGuardia Community College in Queens, New York. The school opened in 1984 with a mission statement that clearly set guidelines on how to educate a limited-English-speaking population. Teachers are expected to pay attention to discourse and communication as they present the content associated with mathematics and other curricular areas. Teachers become observers of the learning process as it unfolds and learn to approach their subject from the point of view of the student.

Dealing with Word Problems

The International High School has thirty-nine language groups, multiple levels of English proficiency, and different mathematics achievement levels. Students who lack English language proficiency, particularly in the area of reading, often find that word problems are obstacles in learning mathematics. To help them in this area, teachers encourage students to create, write, and solve their own word problems. One technique is to have groups of students select strips of paper from paper bags labeled "characters," "theme," and "plot." Descriptive words and phrases are written on these strips. A strip from the "character" bag might say "supermarket cashier," "grocery shopper," "manager," or "butcher." The students are asked to think of possibilities within a given situation and to create a word problem. For example, they might conceive of a word problem that involves finding the price of an item after a 10 percent discount and act it out using the roles of customer and cashier. This creates an opportunity for the students to explore language in a mathematical context by using language skills to discuss and create a word problem.

Another technique is to remove the question from word problems. This forces the students to focus on the stated relationships in the problem, predict a conclusion, and then create a question that leads to a solution. For example, the teacher might remove the last sentence from the following word problem: "The Mets lead the Giants by three runs. The Giants have two runs. *How many runs do the Mets have?*" This leads to lively discussions because the students often have to defend their conclusions to their classmates. The variety of problems that the students create and the manner in which they justify them give the teacher an idea of what the students have mastered. This activity is thus useful at the end of the term as an informal assessment tool.

Collaborative Learning

At the beginning of the school year, rules are established for group work in the mathematics class. Different language groups are encouraged to sit together in order to encourage the use of English. To engage students in the process of collaboration, the teacher searches for an introductory problem that allows students to discuss and explore the possibilities of an open-ended problem. One such problem is the "bridge problem," which is part of the New York City Board of Education Sequential Mathematics I Curriculum Guide. This problem depicts four people with a lantern who wish to cross a bridge in the middle of a jungle in the dead of night. Information regarding the constraints on crossing are given to the students, but the precise question that would initiate a particular solution plan is withheld. Thus, the problem stimulates class discussion, models collaboration, and allows students to experiment with a number of possible questions that might be solved given the information provided. The teacher reads the problem aloud to the whole class and answers any questions that the students have. The teacher often draws on the chalkboard a picture of a bridge and a lantern for clarity. Working in groups, the students read and search for a solution. After a few minutes, the teacher stops the class and asks for feedback. A common comment is, "You can't solve it because there is no question" to which the teacher replies, "Make your own." The students' questions are written on the chalkboard and are read, discussed, and ranked according to their relevance to the problem. The students are encouraged to present their solutions to the class and receive points for doing so. By the end of the cycle, most students have made an attempt to present different solutions to the problem.

Putting All Multiple Approaches Together

In designing these activities, the teacher must focus on the mathematics curriculum and study how the concepts are organized so that the focus is not just a sequence format. To bridge deficiencies and maximize learning, word problems that can be solved in a variety of ways are presented. For example, the class was able to solve one problem using arithmetic. The class was then challenged to solve it using algebra. Over the course of several days, the students presented several types of solutions; one group solved it using simultaneous equations. The students gained a flexibility in realizing that there is not just one prescribed method for solving a given problem but that several are possible and that this makes mathematics an interesting and dynamic field. Some types of problems that can be used in this manner are long distance telephone call problems, parking fee problems, and postal rate problems. The example below shows how a telephone call problem can be modeled arithmetically, algebraically, or as a step function:

Long distance telephone call problem: What is the cost of a five-minute phone call if the first minute costs \$0.45 and each additional minute costs \$0.15?

Minute	Cost of Call
1	0.45
2	0.45 + 0.15 = 0.60
3	0.45 + 2(0.15) = 0.75
4	0.45 + 3(0.15) = 0.90
5	0.45 + 4(0.15) = 1.05

Students can make predictions and generalizations and thereby discover a pattern. From this pattern they can derive a formula and work the problem either through arithmetic (cost of call = cost of first minute + cost of each additional minute) or through an algebraic formula (if x = cost of call and n = number of additional minutes, then $x = \$0.45 + n(0.15)$). The teacher can also model a different way of deriving an answer using a step function such as the following:

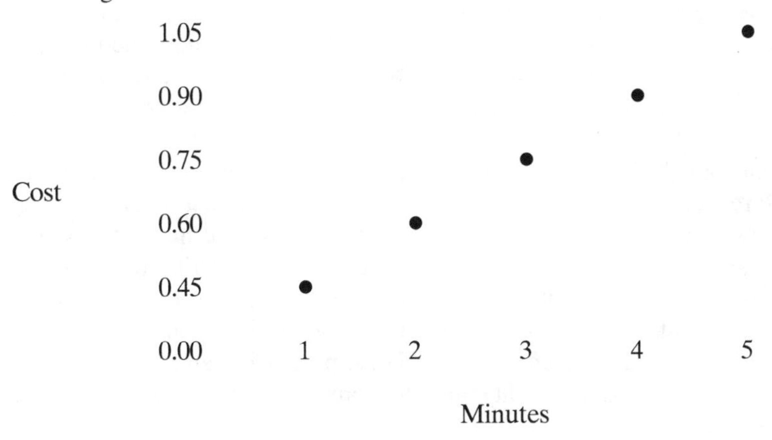

Collaboration in problem solving allows students to share their approaches. Students who need remediation in mathematics often do not know why things are done in a particular way. Such students often rely on memorization. As students move further in their studies, they need a new set of thinking skills: analysis, decision making, and synthesis. It is important for mathematics teachers to be aware that LEP students need to make sense of mathematics.

Student collaboration provides an opportunity to observe how students learn. They need to talk, read, write, and explain to others in order to reach understandings. LEP students need constant integration and reinforcement of skills—and not just drills, which are usually done in isolation, without a meaningful context. When the teaching approach incorporates collaboration coupled with reading and writing, students begin to make meaningful connections, which are vital to the process of learning and recall.

Student Journals

Journals can be very helpful in getting reluctant students to participate in class. A teacher cannot take anything for granted when working with students. Having them listen to lectures is not enough—students need to share what has been understood. For example, in a sequential mathematics class, the instructor followed the prescribed curriculum closely. A Korean student was having difficulty. After a discussion with the student, the teacher became aware that he had not understood the topics that had been presented but was reluctant to let the teacher or his classmates know. The teacher began to use journals to create a silent conversation between herself and her students. This enabled the students to discuss freely the areas of difficulty and to raise questions. The instructor was able to use information gained from the journal entries to pace instruction, that is to know which subjects needed more or less attention. Instead of drill sheets, the instructor began to pose questions so the students could make sense of what was being discussed in class. It was not enough for the students to say "I don't understand." They had to articulate the problem to the best of their ability. This information became useful before a test to focus on the test review and after the test to measure how students felt about it.

Journals can also be used to summarize what was understood. Students are asked to reflect on what happened in class, what they understood, and, to the extent possible, how the concepts they studied are useful in everyday life. Students are expected to observe their surroundings from a different point of view. When they write about abstract mathematical concepts, they often refer to everyday events. For example, they can be asked to reflect on such concepts as order, the commutative and the associative properties, and the difference between equality and inequality and to compare these to activities and concepts in everyday life.

Journal writing allows students to reflect on what they are learning, extend ideas, discuss solutions and strategies, and, most important, create meaning for themselves. To bridge the awkward constructions found in mathematics, students are asked, "How can you state this in your own words?" or "Rewrite this in your own words." This invites the students to express their knowledge and organize and sequence their ideas in their own way. For example, the teacher can ask groups of students to sequence in writing the solution to a problem. They are asked to write what they would do first, next, and so on, until the directions for attaining a solution are recorded. Groups then exchange papers and solve the problem using the directions provided by their classmates. This makes students aware of their own deficiencies and gives them feedback on how to remedy them.

Classes that use these techniques are noisy! At the beginning of each cycle, the teacher writes a letter to the students describing how they will learn mathematics. They are told that for mathematics to make sense, it must be explained, written about, and thought about. They are urged to discuss how problems are solved and to object in a polite way if they find that they disagree with their classmates or the teacher.

Assessment

This teaching method creates a problem when it is time for assessment: Once students get into the habit of collaborating, they find it difficult to work in isolation at test time. There are three ways to overcome this problem. The teacher can give variant versions of the test so that students sitting together do not share the same version, give group tests, or do both. On Monday, students might take a group test where the teacher gives each table a grade, and each student in a group gets the same grade as the group. Students work individually at first, then they work as a group in which they discuss procedures and strategies for solving problems. These tests are generally more difficult than individualized tests and consist of more parts. Geometry problems work very well. When students fail to collaborate, the teacher can intervene as a mediator. On Friday, students take an independent test. Both test grades count toward the final grade.

Test taking is a learned process and students need to see it as such. To discourage copying, students are told to inform the teacher when they get stuck. Then, they can draw a box around the problem and discuss the solution with someone in their group. The name of the helper is written inside the box, and this individual gets one-half credit for the problem. This technique helps to foster collaboration and is useful in getting students "over the hump" when they encounter difficulties.

Vocabulary is learned within the context of the lesson. When a new lesson is presented, the class focuses on the vocabulary. Whenever possible, the instructor stresses roots, prefixes, and suffixes. Together, the class brainstorms and makes lists of new words. The students are asked what they notice about the meanings of the new words. They play word games using letter tiles such as those found in Scrabble. The teacher gives the textbook meaning of particular mathematical terms orally in a sentence, and students try to guess the terms and arrange the tiles to spell them. The winning group gets five points, which can be applied to a test.

All students need to experience mathematical concepts presented from the concrete to the abstract, even in high school. In classrooms where students are heterogeneously grouped, this presentation is essential for success. In most situations, teachers approach a class by addressing the needs of the average student in the class. The high achievers and the low achievers tend to suffer.

By presenting concepts through hands-on activities, which move from the concrete to the abstract, the teacher can reach all students.

For example, teachers at the International High School link mathematics, physics, literature, and physical education through the common theme of motion. In mathematics and physics classes, students are engaged in activity lessons that require working, discussing, and interacting in groups, writing observations, drawing conclusions, and making predictions. As lessons are completed, groups make oral reports to their classmates and field questions, which often call for further clarification. Learning is supported through such discursive tools as peer intervention, manipulatives, and calculators and computers.

Mathematics teachers often forget that mathematics is a tool for the sciences. In the motion class, mathematics skills are used as a tool to explain and clarify physics concepts. Students are exposed to skills not normally found in the prescribed mathematics sequence and learn mathematics skills in a meaningful context.

Another dimension of the motion class is the assessment of students through the use of a portfolio. Students keep written work in folders in each of their classes. Twice in the cycle, teachers announce that students must begin to create a portfolio of their work. They are given basic questions to guide their writing. Students write about their personal goals; what motion means to them; what is common to their mathematics, science, literature, and physical education classes; Newton's laws; and what they feel they have achieved.

As an assessment tool, the portfolio presents a total picture of an individual's work in class. The portfolio represents what has been done over time and what has been learned by the student. Unlike a test, the portfolio has a broad focus: Students must reflect on each completed task to choose which is representative of their best work. Furthermore, students actively learn as they assemble and consolidate their activities. When they begin to make sense, in writing, of all their work, they begin to make connections and to extend ideas.

Completed portfolios are shared with the class. First, students grade their own work. Then, they ask two classmates to read, comment on, and grade the work. Students then choose two faculty members to do the same. The final step is for the writer to sit down with the two classmates and two teachers. The writer describes the work done and the readers react by stating what they feel to be the main achievements.

Despite differences in English proficiency, all students are able to discuss Newton's laws, give everyday examples that illustrate the concept of motion, discuss the connections among the four classes, and present what has been learned. For LEP students, portfolios reveal what has been learned and make students responsible for presenting this information to the teacher and to the class. Learning in this setting is thus a public activity.

Collaboration in mathematics is facilitated when the school structure supports and continuously models the process. At the International High School, teachers learn about collaboration from their fellow faculty members. The mathematics curriculum includes talking, sharing, and writing to lessen the effects of different language and ability levels in the classroom. This approach provides an opportunity to teach LEP students in a way that does not view their language difficulty as a handicap but as a temporary condition that can be overcome. Working in this setting, teachers often take different paths. Some teachers have developed a hands-on approach to teaching mathematics; others have developed a more traditional approach. All teachers nevertheless use second-language learning techniques to facilitate mathematics learning. If at the end the students are on the same road, the journey is worthwhile.

Resources for Implementing the Communicative Approach

A number of resources that will help teachers and administrators implement communicative mathematics teaching are described briefly below. Also listed are the names, addresses, and phone numbers of schools and organizations that can be contacted to order the resources or to get further information. Full bibliographic information is given in the list of references.

Language Development through Content: Mathematics Book A (Chamot & O'Malley 1988) is published as a student's book and a teacher's guide. There are seven units dealing with number and place value, addition, subtraction, multiplication, division, four operations (integration of all four operations), and fractions. Each lesson is accompanied by language objectives, mathematics objectives, and learning strategies and is presented on the basis of a five-phase lesson plan (preparation, presentation, practice, evaluation, and explanation). The materials can be used in grades 4–10 either in mainstream general mathematics classes with high concentrations of LEP students or in English as a Second Language (ESL)/mathematics classes. They are currently being used in a grade 4–10 ESL/mathematics curriculum developed by the Arlington County (Virginia) Public Schools (1990). The materials and the approach are based on the "CALLA (Cognitive Academic Language Learning) approach" to language and content learning developed by Chamot and O'Malley (1986).

CALLA in Action (Arlington County Public Schools 1991) is a videotape that includes segments from CALLA mathematics classes, science classes, social studies classes, and literature classes taught by teachers in the Arlington County public schools. It is available through the Arlington County Public Schools.

English Skills for Algebra (Crandall, Dale, Rhodes, and Spanos 1987) contains five units of supplementary activities for use in intermediate and high school prealgebra and beginning algebra classes. Units 1–4 are published as student books and tutor books. This format allows students to practice in pairs, with one student acting as the tutor and the other as the tutee. The units are infused with language learning activities involving speaking, reading, writing, and listening. In unit 3, there are exercises that allow students to work together to draw diagrams and pictures corresponding to word problems or, conversely, to write word problems corresponding to pictures and diagrams. In unit 4, students are encouraged to discuss paraphrases and inferences of selected mathematical definitions and theorems. Unit 5 is a glossary that provides information pertaining to the meaning and application of mathematical terms and phrases. The use of these materials is fully described in a videotape training package developed by the Miami-Dade Community College (1989).

The Pre-Algebra Lexicon (PAL) (Hayden and Cuevas 1990) contains a structured listing of mathematical terms and expressions most commonly found in prealgebra courses and textbooks. In addition to the listing, the PAL provides annotations that are succinct analyses of many terms in the structured listing. Another section presents diagnostic assessment techniques that provide teachers with strategies through which the development of mathematical language skills can be assessed within the context of daily instruction. The techniques are categorized according to four mathematics categories (concepts, operations, word problems, and problem solving) and four language skills (listening, speaking, reading, and writing).

The assessment techniques in the *Pre-Algebra Lexicon* are based in part on the techniques provided in *Assessment Alternatives in Mathematics* (Stenmark 1989). The reader should consult the latter for useful information on student portfolios, writing activities, open-ended questions, interviews, and question formatting. The sections on open-ended questions (pp. 16–19) are particularly illuminating for teachers interested in focusing on communication. Teachers are encouraged, for example, to have students consider three statements that might be true of the following situation: "A friend says he is thinking of a number. When 100 is divided by the number, the answer is between 1 and 2" (p. 16). The value of the exercise is that students are invited to verbalize their conceptions and misconceptions about mathematics and teachers are able to diagnose where students need assistance.

Communicative Math and Science Teaching: A Video (Center for Applied Linguistics 1989) presents classroom examples of teachers and students engaged in activities that emphasize communication, cooperative learning, peer tutoring, and the uses of games. The videotape is accompanied by an instructional guide (Spanos 1990), which briefly presents the communicative approach to mathematics teaching and provides discussion activities for teacher workshops. The mathematics classes were taped at the International

High School at LaGuardia Community College, the Bronx High School of Science, and Lanier Intermediate School in Fairfax, Virginia.

"Integrating Mathematics and Language Learning" (Dale and Cuevas, 1992) discusses the role of language learning in mathematics, learning mathematics through a second language, and learning a second language through mathematics. Instructional strategies are presented for use both in mathematics classes and in ESL classes. Specific examples are drawn from the instructional materials developed by Crandall, Dale, Rhodes, and Spanos (1987).

"Linguistic Features of Mathematical Problem Solving" (Spanos, Rhodes, Dale, and Crandall 1988) discusses the rationale behind the communicative approach to mathematics teaching and learning. The article contains a discussion and a categorization of some problematic features of mathematics language and is based on transcriptions of student problem-solving sessions.

Innovative Strategies for Teaching Mathematics to Limited English Proficient Students (Secada, Carey, and Schlicher 1989) is a guide for teaching LEP students academic skills in mathematics. The guide furnishes sample mathematics lessons and suggestions for implementing cognitively guided instruction and active mathematics teaching. The lessons are modeled after those currently being used in an ESL mathematics curriculum developed by teachers in the Alexandria City (Virginia) Public Schools (1988).

For information and guidance pertaining to the use of writing in the mathematics classroom, the reader should consult *Writing to Learn Mathematics and Science* (Connolly and Vilardi 1989) and *Using Writing to Teach Mathematics* (Sterrett 1990).

In addition to the resources identified above, the reader should not fail to consult the *Curriculum and Evaluation Standards for School Mathematics* (NCTM 1989) and *Professional Standards for Teaching Mathematics* (NCTM 1991) for numerous suggestions and examples of classrooms that incorporate communication activities.

Contact Schools and Organizations

Alexandria City Public Schools, ESL Center, 3801 West Braddock Road, Alexandria, VA 22302, (703) 824-6660.

Arlington County Public Schools, HILT/Special Needs/CALLA Office, Washington-Lee High School, Room 12-E, 1300 North Quincy Street, Arlington, VA 22207, (703) 358-6233.

Bard College, Institute for Writing and Thinking, Annandale-on-Hudson, NY 22504, (914) 758-7431.

Center for Applied Linguistics, 1118 Twenty-second Street, NW, Washington, DC 20037, (202) 429-9292.

International High School, LaGuardia Community College, 31-10 Thompson Avenue, Long Island City, NY 11101, (718) 482-5456.

ERIC Clearinghouse on Languages and Linguistics, 1118 Twenty-second Street, NW, Washington, DC 20037, (202) 429-9292.

National Clearinghouse for Bilingual Education, 1118 Twenty-second Street, NW, Washington, DC 20037, (800) 321-NCBE.

National Council of Teachers of Mathematics, 1906 Association Drive, Reston, VA 22091, (703) 620-9840.

Upper Great Lakes Multifunctional Resource Center, University of Wisconsin, 225 North Mill, Madison, WI 53706, (608) 263-4220.

References

Alexandria City Public Schools. *ESL Math 3 Curriculum.* Alexandria, Va.: Alexandria City Public Schools, 1988.

Arlington County Public Schools. *CALLA in Action: A Video.* Arlington, Va.: Arlington County Public Schools, 1991.

________. *HILT Mathematics Curriculum Guide.* Arlington, Va.: Arlington County Public Schools, 1990.

Center for Applied Linguistics. *Communicative Math and Science Teaching: A Video.* Washington, D.C.: Center for Applied Linguistics, 1990.

Chamot, Anna Uhl, and J. Michael O'Malley. *A Cognitive Academic Language Learning Approach: An ESL Content-based Curriculum.* Washington, D.C.: National Clearinghouse for Bilingual Education, 1986.

________. *Language Development through Content: Mathematics Book* A. Reading, Mass.: Addison-Wesley Publishing Co., 1988.

Cocking, Rodney R., and Jose P. Mestre, eds. *Linguistic and Cultural Influences on Learning Mathematics,* Hillsdale, N.J.: Lawrence Erlbaum Associates, 1988.

Connolly, Paul, and Teresa Vilardi, eds. *Writing to Learn Mathematics and Science.* New York: Teachers College Press, 1989.

Crandall, JoAnn, Theresa Corasaniti Dale, Nancy C. Rhodes, and George Spanos. *English Skills for Algebra.* Englewood Cliffs, N.J.: Prentice Hall Press, 1987.

Cuevas, Gilberto J. "Mathematics Learning in English as a Second Language." *Journal for Research in Mathematics Education* 15 (March 1984): 134–44.

Dale, Theresa Corasaniti, and Gilberto J. Cuevas. "Integrating Mathematics and Language Learning." In *The Multicultural Classroom: Readings for Content-Area Teachers,* edited by Patricia A. Richard-Amato and Marguerite Ann Snow, pp. 330–48. White Plains, N.Y.: Longman, 1982.

Hayden, Dunstan, and Gilberto J. Cuevas. *The Pre-Algebra Lexicon.* Washington, D.C.: Center for Applied Linguistics, 1990.

Miami-Dade Community College. *A Training Package for English Skills for Algebra.* Miami, Fla.: Miami-Dade Community College, 1989. (Project funded by the U.S. Department of Education, Fund for the Improvement of Postsecondary Education, Grant no. G008730483.)

National Council of Teachers of Mathematics. *Curriculum and Evaluation Standards for School Mathematics.* Reston, Va.: The Council, 1989.

________. *Professional Standards for Teaching Mathematics.* Reston, Va.: The Council, 1991.

Secada, Walter G., and Deborah A. Carey. *Teaching Mathematics with Understanding to Limited English Proficient Students.* ERIC Clearinghouse on Urban Education, Urban Diversity Series, No. 101. New York, N.Y.: ERIC Clearinghouse on Urban Education, 1990.

Secada, Walter G., Deborah A. Carey, and Roberta Schlicher. *Innovative Strategies for Teaching Mathematics to Limited English Proficient Students.* Washington, D.C.: National Clearinghouse for Bilingual Education, Program Information Guide Series, 10, 1989.

Spanos, George. *Communicative Math and Science Teaching: An Instructional Video Guide.* Washington, D.C.: Center for Applied Linguistics, 1990.

Spanos, George, Nancy C. Rhodes, Theresa Corasaniti Dale, and JoAnn Crandall. "Linguistic Features of Mathematical Problem Solving: Insights and Applications." In *Linguistic and Cultural Influences on Learning Mathematics,* edited by Rodney R, Cocking and Jose P. Mestre, pp. 221–40. Hillsdale, N.J.: Lawrence Erlbaum Associates, 1988.

Stenmark, Jean Kerr. *Assessment Alternatives in Mathematics: An Overview of Assessment Techniques That Promote Learning.* Prepared by the EQUALS staff and the Assessment Committee of the California Mathematics Council *Campaign for Mathematics.* Berkeley, Calif.: Regents, University of California, 1989.

Sterrett, Andrew, ed. *Using Writing to Teach Mathematics.* Washington, D.C.: Mathematical Association of America, 1990.

11

Geometry for All Students: Phase-based Instruction

Janet C. Bobango

One of the societal goals set forth in the NCTM *Curriculum and Evaluation Standards for School Mathematics* (1989) is opportunity for all. Mathematics in general is indeed a critical filter with respect to entrance into college and certain major areas of study as well as to career opportunities (1989, p. 4). Geometry in particular provides opportunities for practical applications and problem-solving situations and plays an important supporting role in other mathematics as well as in such fields as engineering, architecture, astronomy, and physics (Niven 1987).

In spite of the important place geometry holds for students' futures, only about one-half of the students in the United States take a formal geometry course. Of those who do, approximately 33 percent are successful in terms of being able to prove theorems and exercises deductively (Senk 1985). Additional information on student success in geometry was reported in the Fourth National Assessment of Educational Progress (Lindquist 1989). Performance on questions relating to geometry and measurement was weak for all students. Furthermore, black and Hispanic students' achievement lagged behind that of other minorities and white students (pp. 140–41). The assessment also found gender differences on the geometry subscale at grade 11, with males scoring significantly higher than females (p. 154). There was also a significant difference in favor of seventeen-year-old males on items requiring higher-level thinking, which was attributed to those questions involving "applications of numbers and operations, measurement, and geometry" (p. 158).

Goals and Standards

The data cited above indicate that student involvement and performance in geometry need to be improved. The National Council of Teachers of Mathematics has supported this aim by including geometry as one of the standards for all grade levels. Success in geometry can contribute to achieving NCTM's goals for all students, which include gaining confidence in their mathematical abilities, becoming problem solvers, communicating mathematically, and reasoning mathematically.

The van Hiele Theory of Geometric Thinking

In searching for an explanation for this lack of involvement and success as well as for guidelines that will help solve the problems, researchers have taken note of a theory that incorporates many of the goals expressed in the *Standards.* The van Hiele theory of geometric thinking (van Hiele 1986) claims that students are not ready to do deductive proofs until they have progressed through three previous levels:

1. Visualization—students pay attention to the whole figure and refer to it by name.
2. Analysis—students focus on the parts of configurations, such as the relationships of sides and angles.
3. Informal deduction—students can define terms, understand if-then statements, form subclasses of configurations, and follow informal deductive proofs.

There is a data base furnishing evidence that students who enter a formal geometry course at the third level do in fact have a greater chance of success (Usiskin 1982). One of our jobs as mathematics educators, then, is to meet two needs: (1) provide students with backgrounds that will enable them to achieve level 3 before they enroll in a traditional geometry course at some time during grades 9–12, and (2) build students confidence in their abilities to succeed in geometry.

To enable students to progress to a higher level requires five phases of instruction. The first phase, *information,* gives students the opportunity to discover what they will be studying and which experiences and concepts from their backgrounds will be useful. It also gives teachers some idea of the conceptual understandings and knowledge their students already have about the upcoming areas of study. The next two phases involve two kinds of tasks. One kind falls into the phase van Hiele called *guided orientation.* In this phase activities are relatively easy and the outcomes are anticipated. The second kind is classified under the phase *free orientation* and is more complex, perhaps having more than one solution and certainly more than one way of approach; students are left to their own resources to figure ways to complete the activity.

Following either kind of activity, students should be given time to communicate their approaches and findings to each other. In this phase, called *explicitation,* students should be permitted to use nonstandard terms during discussions because basic to the theory is that the vocabulary used at different levels is unique. Students need the opportunity to assimilate concepts in a manner meaningful to them, without having to focus on terminology. The final phase, *integration,* is similar to review and is a time for students to summarize what they have learned as well as a time for teachers to identify ideas that have not been understood by all students.

Contrary to the levels, the five phases are not necessarily sequential. If the integration phase reveals that students do indeed need more work, teachers should use their judgment in returning to the appropriate phase or phases.

Several books on informal geometry, many manipulatives, and articles in mathematics education journals are available to help in planning for geometric instruction to reach all students. Some of these materials were used to formulate the course discussed below, which was designed in the spirit of the *Standards* and based on the van Hiele theory of geometric thinking.

Demographic and Contextual Data

This special semester-long course, designed with phase-based instruction, was implemented with a class of twenty-four students who had studied traditional algebra 1 material for three semesters. The class was divided equally between males and females. Of the twelve females, one was African American and one was Hispanic. The students in this suburban-rural school ranged from grades 9 through 12; consequently, they had different backgrounds in mathematics.

In spite of varying mathematical experiences, one of the major goals of the course was to give students opportunities to succeed. The intent was that success would increase students' confidence in their abilities to do mathematics, would promote a positive attitude toward mathematics, and would encourage them to elect to continue studying mathematics. In particular, those students electing to take formal geometry would furnish some evidence that this course, grounded in phase-based instruction, had been successful.

Description of the Course

Basically the course was divided into five broad categories: triangles, including angle measurement and congruent triangles (27 days); quadrilaterals (19 days); parallel and intersecting lines (13 days); area and perimeter (8 days); and circles (8 days). Naturally, things overlapped, and the teacher made an effort to help students integrate the concepts that had been studied before. Students knew that they were undertaking something that was somewhat experimental, that they would not be using a textbook, that their classwork was very important, and that there would not be much homework.

It was explained very clearly to students that their grades would be determined in the following way: tests and quizzes, classwork, and projects would each be worth 30 percent toward their quarter marks. Homework and journal entries would constitute the remaining 10 percent. Initially students were paired with a partner for some activities. Partners were expected to help each other not only with certain activities but also in understanding concepts as well as explaining what was done any time the other was absent. This was especially necessary because no textbook was being used. Later in the quarter, students gave the teacher the name of the person with whom they would like to work. This input was used to form groups of four for the remainder of the semester. In order to give the reader some idea of the course content, a description of selected lessons from the triangle section follows. A similar pattern was used for the other categories.

Information Phase

The first class meeting was designed to give the teacher some idea of what the students already knew about triangles and to help the students realize what they already had in their backgrounds that would help them in the upcoming unit. Using an activity similar to the one from Burger and Shaughnessy's work (1986), the teacher gave students time to identify triangles on a sheet of paper. (See fig. 11.1.)

Then students were asked to express their ideas of what was true about a triangle. The teacher accepted all answers, writing them on the board. Ideas included such things as three sides, can have 90° angle, straight lines, has space, connect, perfect, same size, and three points.

During the discussion of the numbered configurations in figure 11.1, students had no difficulty in agreeing that #1, #7, and #9 were triangles. There was substantial disagreement, however, about #3, with some students calling it a "fat triangle." In #5 and #14, students suggested that others ought to use their imaginations to complete the triangles. One student in particular concentrated on three points as a major factor in deciding if a figure was a triangle. This made #3, #6, #8, and #13 triangles in her eyes. Triangles #6 and #13 were just "smashed in," and if "blown up," they would be perfect triangles. Students definitely had the notion that there is something called a "perfect triangle."

As the discussion continued, students started to become impatient, as evidenced by such comments as "Let's settle this thing" and "Just tell us." However, they were willing to express their ideas and did come to a consensus by the end of the class and wondered what else they would be doing.

During the remaining few minutes in the period, the students were asked to react to the class activity in their journals by summarizing what they had learned as well as candidly commenting on how well they liked it. Some responses confirmed the impatience noted above and indicated students expected to be given information and were not accustomed to being asked to participate in this type of discussion. Several comments were similar to the following:

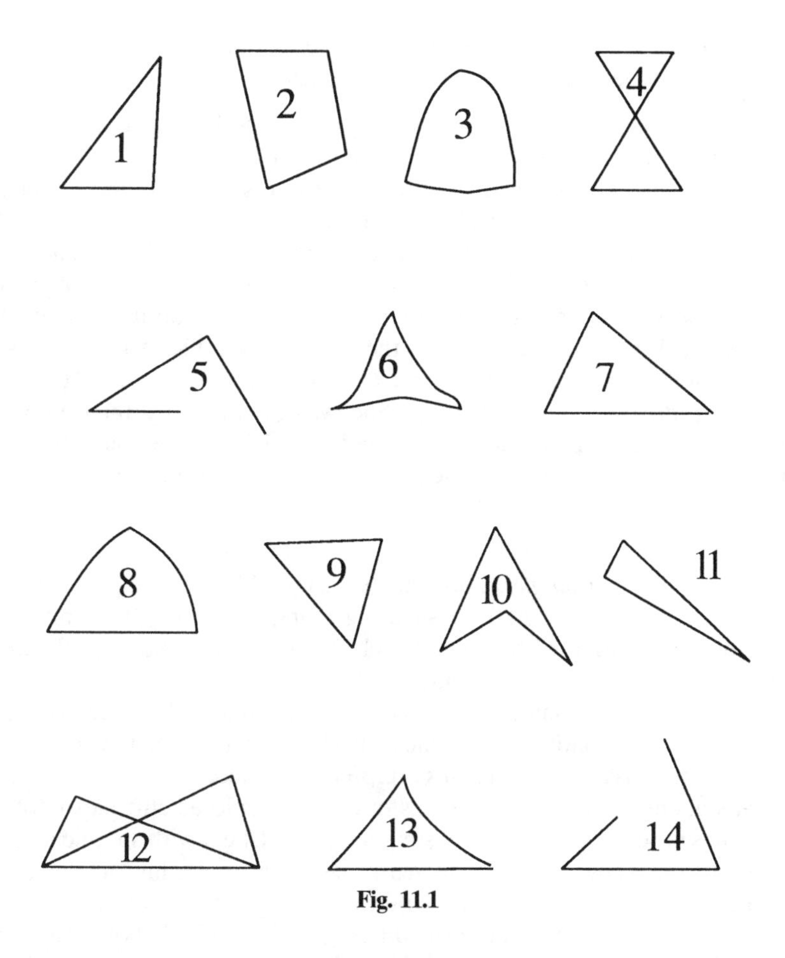

Fig. 11.1

- "Today was kind of boring because I learned it last year. I was annoyed that people did not know this already."
- "I thought all the arguing was ridiculous. She should have just told us."

But an equal number of entries indicated a liking for the approach:

- "I don't think it was stupid, even though it took a while to get through the exercise. I feel I won't forget the rules."
- "What we did today was all right. It made people have to think about triangles."
- "I never knew that a triangle had to have straight lines. It wasn't too bad. We got to all work together."

One final comment perhaps summarizes the overall assessment of this information phase and illustrates the mixed reactions students experience:

- "I thought it was kind of stupid because you should have just told us. But it did help me learn a little better, so I guess it wasn't too stupid."

Recall that the purpose of the information phase is to give students an indication of what they would be doing and what they could use from their past experiences. This seemed to have been successful. In addition, though, the teacher is supposed to gain insight into where the students are with respect to a topic. Therefore, following the class, the teacher was asked if she had been able to tell from the discussion who knew what a triangle is. She thought so, although she was not certain which students were still confused about the curved sides. When concentrating on the vocabulary students had used, she recalled *point, lines, connecting, complete triangle, perfect triangle, 90-degree angle, long stretched out,* and *space.* She was surprised by the confusion students had in identifying the triangles in figure 11.1, and she found this part of phase-based instruction to be helpful in planning future lessons.

Classwork

Activity—making triangles. This activity was used because it gave students another opportunity to work on visualizing triangles (level 1), to think more about the composition of triangles (level 2), and to gather data and subsequently make conjectures regarding the lengths of sides needed to make a triangle (level 2). The phase of this activity is fundamentally guided orientation because it is relatively easy, the teacher provides direction in how to accomplish the task, and the results are predetermined.

Each student received eleven strips of different colored stiff paper (other possibilities are pipe cleaners or D-stix) and a worksheet with specific lengths in centimeters. They were told to mark each strip according to the given measures and to fold on the marks to form a triangle; there was a little extra tab for gluing on the ones that formed triangles. It required about thirty-five minutes for students to complete the eight that were possible. The remaining fifteen minutes were spent having students discuss findings (explicitation phase) and write in their journals.

During the work time, there were times when the room was quiet and there were times when students were conversing. Sometimes they were checking to see what a particular colored triangle looked like or checking to see which strips did not form triangles. But there was also an announcement regarding a possible objective for the day: "I think we're trying to learn if a triangle can be any length." Some students still wanted to have the teacher tell them the correct answer. She made it clear that *they* were going to do the "telling." As in most classrooms, students finished at different times. When the first students appeared to be finished, the teacher reminded them to make a combination with one that would work and one that would not and to have an explanation ready. Also, they could write in their journals.

During the explicitation phase, it became clear that students thought they had to call one side a base. Some had the impression that *the* base was the longest side. Because students had triangles in front of them, they could rotate them, observing that the triangle had remained unchanged. This enabled the teacher to emphasize that orientation did not make a difference. (This part of the discussion indicated that many of the students were at the visualization level.) Level 1 students want the "base" to be parallel to the bottom of the page and think that after rotating it so the triangle "points down," a different triangle has been formed. The discussion continued with students conjecturing that the sum of two sides had to be equal to or greater than the third. Another countered that for the 10, 5, 5 strip that was not true, so it was agreed to remove the "equal" condition. One student recalled classwork the day before when they had taken three numbers and made possible combinations of pairs and used that information to check the lengths of sides needed to form a triangle. The discussion ended by applying the class's conclusion—that of having to compare the sums of all pairs to the length of the remaining side—to a student-generated example of 3, 2, 1.

In summary, this activity was a lot of work for the teacher, and it created some confusion during class—rows were not straight and, yes, there was even some talk that did not pertain to geometry. For the most part, however, students were actively involved in the task, they were communicating mathematically, and there was cooperation in arriving at the appropriate conclusion. The students had succeeded. The teacher did not just tell them or write the theorem on the board.

Because all the students were involved and the discussion flowed easily, it seemed as though the objective had been achieved by all students. The students journal entries provided a means other than testing to check for their reactions and for understanding by all students. This activity received positive comments from all students along the lines of less boring, fun, helped us better understand, different, keeps students attentive, and enjoyment "because I actually figured something out in my head that has numbers!" Furthermore, except for one, all summaries of what they had learned were accurate and similar to this one, "If two numbers add to be bigger than the leftover number in any combo, it's a triangle." What was that one lone comment? "I really don't know what I learned. I guess nothing for today." Perhaps we never reach all students in a class at any single given time.

Activity—measuring sides of triangles. The next day, class was begun with a review that indicated students had made progress. The fight over whether sides could be curved had disappeared. Students had accurate perceptions of what triangles look like and were focusing on their properties. A review of the concept about lengths for those who had been absent or had not understood came from interaction between two students.

Following the review, students were given envelopes containing seven different triangles whose measurements were the same as the previous day's triangles. They were to measure the sides to see which triangles "matched" and which ones were alike in two ways or one way, as well as describe how they were different. During the discussion it was evident that students knew they could look at triangles in many different ways and that sometimes they needed to flip them in order to put one on top of the other. (Later in the course, this exercise was referred to when students were working on the different ways to show triangle congruence.)

Journal entries once again gave all students the opportunity to communicate mathematically and did indicate that most had understood the intent of the activity. There was evidence that students knew that a triangle would not change simply because it was rotated or flipped:

- "Today we learned to look at triangles differently—we have to learn to flip them over, because the same triangle could look different but just be flipped!"
- "I learned today that there are many comparisons between triangles. All you have to do is look for them."

Activity—measuring angles. The students had done some work with protractors, and the teacher thought that they could measure acute, obtuse, and right angles as well as have a sense of what they looked like approximately. An exercise, with partners helping each other, revealed, however, that students were not good at measuring angles. Gross errors were made either because of carelessness or misconception. Therefore, another exercise was used that required students to measure angles. In addition, this exercise focused on linear pairs of angles forming a straight line segment. Students were given large cutout angles that would form a straight line segment when joined together. They then measured their angles and recorded the data on the board. One student knew they were supposed to add to 180, so students remeasured and changed some of their original work. This activity really enabled students to help each other in using protractors, in turning angles around, and in noting that those that fit together consisted of one acute and one obtuse angle or two right angles.

Again, their journal entries indicated that the students thought they now knew how to measure angles. The entries also, however, showed some incorrect usage of standard terms such as *triangle* for *angle* or "two degrees add up to 180 make a straight line." This seems to be an indication that students were at least in transition from level 1 to level 2 but probably not quite there.

Computer lab work. In keeping with NCTM's call for students to have access to computers for individual and group work, the class was able to work in the computer lab two times during the triangle unit and a few more times during the other units. Working in pairs, students spent time exploring the program Geometric Supposer: Triangles and generating obtuse, acute, and right triangles. They gathered data about the angles and lengths of sides, then used the information to try to write definitions. In general, students liked using computers, they thought computers helped them to visualize the different types of triangles, and they stayed on task, often conversing with their partners about their findings.

Students seemed willing to talk freely about their results and to ask questions. For example, one student asked why it was not possible to have a triangle with two obtuse angles. A clear explanation indicating sound mathematical reasoning was given by another student. Because the classroom climate had become quite relaxed and students accepted the fact that they were not going to be told everything, it was easy for the teacher to hear the type of vocabulary being used and to find out what some of the misconceptions were. Students were now relating angle measures and considering the lengths of sides to combine names for triangles (level 2 behavior). However, some nonstandard vocabulary, such as *corners* instead of *vertices*, was still used by some. In addition, confusion about sides and angle measure surfaced. Someone had asked if it was possible to have a triangle whose angles measure 90, 89, and 1. A second student said it couldn't be because when two were added together, the sum was not greater than the third number. Without this free exchange between students, the teacher may never have been aware of the need to remind students that the conclusion about the sum of two numbers compared to the third refers to the measures of sides that form triangles and not to the measures of their angles.

Projects

During this unit, students had to do one of three projects (two had variations for earning different grades):

1. Find ten pictures of triangles in magazines or newspapers, cut them out, and present them in some interesting fashion. (The most a student could earn for this project was a C.)
2. Using a 5 x 5 geoboard, find triangles that are different and draw them on dot paper.

   30 different ones = C

   35 different ones = B

   40 different ones = A

   Note: Students were permitted to check out geoboards for use in study hall and at home.
3. Given the measures 3, 4, 5, 6, 8, 10, 12 representing the lengths of sides of triangles, list the possible combinations that form triangles.

40 triangles = C

50 triangles = B

50 triangles plus 10 models (using toothpicks, Popsicle sticks, or paper strips) = A

In general, students seemed to enjoy doing projects throughout the semester. During the other units students did other projects, such as making posters representing geometry in the world, making mobiles out of circles containing geometric configurations, and using a specified number of tangram pieces to form squares, trapezoids, triangles, rectangles, and parallelograms. Occasionally some students chose not to do a project, resulting in a failing grade; other students indicated this was their favorite part of the course.

Evidence of Success

In addition to the tests, projects, daily classwork, and journal entries, data were gathered from individual interviews and test scores from the standardized test *Cooperative Mathematics Tests: Geometry* before and after the phase-based instruction. Using an adaptation of Burger and Shaughnessy's (1986) interview process before the course indicated that most students were at level 1 or 0. Students were asked to participate in three activities. One was to draw as many triangles as possible that were different in some way from one another. Then they were given a sheet of paper containing many different configurations similar to those found in figure 11.1. Students were asked to label the configurations and explain why they thought each was or was not a triangle. Finally, they were given a set of eight cutout triangles and asked to group them, explaining what their sets had in common. In responding to questions on the activities during interviews, students tended to use such nonstandard vocabulary as *shorter triangles, longer sides, taller, skinnier, fatter, narrow*, and *slanted sides*. Very seldom did a student use such terminology as *right angle* or *isosceles triangle.* They had a tendency to focus on three "points" with no attention to whether the sides of the figures were curved or straight.

During the postinterview, students were asked to sort the cutout triangles once again. Attention still was more focused on sides than on angles, but students did group them by type of triangle rather than "fatter" and "taller." The rest of the postinterview consisted of the same three activities but with quadrilaterals instead of triangles. When asked to draw four-sided figures, more students used straightedges and drew by type than had done so during the preinterview. Most, but not all, were able to identify correctly squares, rectangles, parallelograms, and rhombuses. Many labeled the sheet of quadrilaterals with more than one name, which clearly indicated that they had gained an understanding of class inclusion (level 3 behavior). The majority of students still struggled with whether a square was a rectangle or a rectangle a square, relying on memory rather than mathematical reasoning. There was, however, definitely more precision and use of standard terminology during the postinterview activities.

In addition to these positive findings, the results on the standardized test measuring content knowledge were favorable. The mean for the pretest was 12.47; for the posttest it was 17.88. This improvement was significant at the .001 level.

The success of the course in terms of students electing to take a formal geometry course was less positive than the other measures. Of the twenty-four students in the course, three graduated and one moved, all males. Of the remaining twenty people, thirteen (seven of the eight males and six of the twelve females) elected to enroll in the formal geometry course. In general, the overall percentage and the number of females enrolling in geometry are quite consistent with national trends, but the percentage of males is higher than expected.

Summary

The phase-based instruction grounded in the van Hiele theory of geometric thinking proved to be helpful in planning lessons and was successful in many ways. Students clearly improved their geometric knowledge and appeared to progress to higher levels. Their attitudes during class were positive, and they were on task throughout most class periods. A substantial percentage (65 percent) of those still in the district also enrolled in the formal geometry course the following year.

The new goals for students set forth in the NCTM's *Curriculum and Evaluation Standards* were central to the phase-based instruction. The activities during the guided and free orientation phases required students to solve problems both independently and in groups. Before, during, and after these activities, students were challenged to reason and communicate mathematically. The activities were chosen with a high potential for students to succeed and consequently gain confidence in their mathematical abilities. Phase-based lessons required students to be actively involved and gave them opportunities to assimilate information as well as construct their own meanings. Long before the end of the course the "just tell me" syndrome had disappeared.

References

Burger, William F., and J. Michael Shaughnessy. "Characterizing the van Hiele Levels of Development in Geometry." *Journal for Research in Mathematics Education* 17 (January 1986): 31–48.

Lindquist, Mary M., ed. *Results from the Fourth Mathematics Assessment of the National Assessment of Educational Progress.* Reston, Va.: National Council of Teachers of Mathematics, 1989.

National Council of Teachers of Mathematics. *Curriculum and Evaluation Standards for School Mathematics.* Reston, Va.: The Council, 1989.

Niven, Ivan. "Can Geometry Survive in the Secondary Curriculum?" In *Learning and Teaching Geometry, K–12,* 1987 Yearbook of the National Council of Teachers of Mathematics, edited by Mary M. Lindquist, pp. 37–46. Reston, Va.: The Council, 1987.

Senk, Sharon L. "How Well Do Students Write Geometry Proofs?" *Mathematics Teacher* 78 (September 1985): 448–56.

Usiskin, Zalman. *Van Hiele Levels and Achievement in Secondary School Geometry.* Chicago: University of Chicago Press, 1982.

van Hiele, Pierre M. *Structure and Insight: A Theory of Mathematics Education.* Orlando, Fla.: Academic Press, 1986.

12

Females, Minorities, and the Physically Handicapped in Mathematics and Science: A Model Program

Camilla A. Heid
Theresa L. Jump

The Center for Urban and Multicultural Education at Indiana University has developed and implemented an award-winning sex equity training project that focuses on junior high and middle school students in the areas of mathematics and science. Project TEAMS (Training for Equitable Attributes in Mathematics and Science) was funded for two years by the Women's Educational Equity Act (WEEA) program and for one year by the Indiana Department of Education's Title II office. The project is unique in developing an eclectic model that combines three sex equity programs: an award-winning teacher interaction program (INTERSECT), a nationally recognized mathematics program (EQUALS), and an outstanding science program (COMETS) (see the end of this chapter for information on these programs). A supplemental science enrichment component for physically handicapped students (SAVI-SELPH) was added to the project to complement the TEAMS model. Few sex equity projects have focused on promoting females who are minority or physically handicapped. Additionally, this project coordinates existing mathematics and science intervention programs while focusing on teacher-student interactions, career education, nonsexist curriculum practices and materials, and strategies for change.

Project TEAMS and its developers were recipients of the 1986 National Science Teachers Association (NSTA)–American Gas Association Science Teaching Achievement Recognition (STAR) Award and the 1987 Lilly Endowment Organizational Creativity in Youth Programming and Commitment to Youth Award in the State of Indiana. The project was honored for its creativity, excellence, and innovative approach.

The purpose of this chapter is threefold. First, this chapter will review the literature on equity based on sex, race, and physical handicaps in mathematics and science and the subsequent impact on the career orientation of junior high and middle school students. Second, the chapter will provide a systematic overview of the Project TEAMS's training intervention model. Finally, the chapter will present anecdotal data collected during the research component of the project.

Rationale

A variety of recommendations for enhancing the United States's dwindling supply of human resources in science and mathematics have surfaced. The most promising solution would be to broaden the pool of qualified professional and skilled workers to include women, minorities, and the physically handicapped, who constitute 60 percent of our population. The traditional pool of trained workers from which we have drawn our scientists, engineers, and technicians is no longer sufficient to meet our current and future needs for employment.

Females, minorities, and the physically handicapped make up the underrepresented population in the trained mathematics and science work force and students in our society. Women are 51 percent of the population and 45 percent of the nation's work force, yet they are only 11 percent of the scientists, mathematicians, and engineers in this country. Women account for 30 percent of the undergraduate degrees in science and engineering but for only 16 percent of the doctorates in the physical sciences and 7 percent of the doctorates in engineering. African Americans make up 12 percent of the U.S. population yet only 2 percent of all employed scientists and engineers. They account for 1 percent of the doctorates in science and engineering. In 1986, eighty-nine African Americans earned the doctorate in science; only fourteen earned that degree in engineering. Although the Hispanic population constitutes 9 percent of the United States population, Hispanics account for only 2 percent of all employed scientists and engineers and for only 1 percent of doctorates in science and engineering. Likewise, 36 million people of working age have some handicap, yet only 94 000 handicapped working scientists and engineers were identified in 1986 (Task Force on Women, Minorities and the Handicapped in Science and Technology 1988).

Why are the physically handicapped, minorities, and females avoiding the fields of mathematics and science? The forces are subtle but quite pervasive. Several factors emerge from the literature as significant deterrents to the pursuit of achievement in mathematics and science by these underrepresented groups.

Early childhood socialization, especially by parents, teaches young children gender-appropriate roles, attitudes, and behaviors. Boys are encouraged to make decisions, explore, fix things, be independent, and learn how things work, whereas girls are taught to be passive, dependent, domestic, and feminine. It is clear that by the time children reach school age, teachers, peers, and significant others exert influence. However, parental values, attitudes, and expectations remain the most dominant forces in shaping and socializing children. Research indicates that parents believe mathematics is a more appropriate activity for males than for females (Fox 1977).

Although there are no measurable differences in mathematics aptitude between males and females in the early grades, elementary school teachers, textbooks, and curricular materials perpetuate the negative gender stereotypes and biases reflected in sex role socialization. Mathematics and science teachers have lower expectations for females, minorities, and the physically handicapped and reinforce science as "male" subjects. Furthermore, research shows that girls have fewer experiences with instruments, materials, labs, techniques, and field trips in mathematics and science (Kahle 1983). White males make up two-thirds of the illustrations in mathematics books, and female role models are almost nonexistent in science textbooks. Illustrations of successful handicapped scientists or mathematicians are totally absent.

An overwhelming number of studies indicate that the early adolescent years of 9–13 are critical for psychological, social, and cognitive development in the fields of mathematics and science. It is during this stage of growth that "goal embedding" takes place (Mandelbaum 1981); peer pressure to conform to traditional sex roles is most intense; children's opinions of themselves and confidence in their abilities are lowered or raised (e.g., female, minority, and physically handicapped students' opinions and confidence change and become lower); and attitudes toward, achievement in, and aspirations in, mathematics and science sharply decline.

Research reports that enrollment in elective science and mathematics courses is closely related to students' perceptions of the usefulness of these subjects to their future lives and their estimation of their ability to be successful in the courses. Among college-intending seniors, the enrollment of African Americans and Hispanic females in four or more years of mathematics or science courses is consistently 15 to 20 percent lower than that of white males (National Research Council 1987). Female, minority, and physically handicapped students select fewer high-level mathematics and science courses; have less interest and confidence in these subjects; develop greater anxieties toward mathematics and science; fail to see the importance of these studies to their future careers and jobs; and believe science to be a masculine domain.

Fear of success has been documented in the literature as being prevalent in females. It has been defined as the "inhibition of achievement orientation by the expectation of negative consequences, especially in male-dominated areas" (Horner 1968). Findings support the "fear of success" in adolescent girls' attitudes toward science and mathematics. "Causal attribution" is another gender-related perception revealed in the literature. The data suggest that boys attribute their successes in their academic coursework to themselves and their failures to external causes, whereas girls reverse this pattern. Girls attribute their success to luck and their failures to their "nontechnological minds."

Females don't start off in schools behind males. In fact, in the early grades, girls stand out ahead of boys in counting and numeration skills and remain on equal achievement levels in mathematics and science throughout the elementary school grades. The National Assessment of Educational Progress (1983) shows no clear pattern of differences in achievement for boys and girls at ages 9 to 13. But as boys and girls progress through junior high and high school, males' average performances exceed those of females at every cognitive level. Also of great importance is the fact that the boys' opinions of themselves and their futures become higher as they move through the educational system but the girls' opinions become lower. "Of the brightest high school graduates who do not go on to college, 70 percent to 90 percent are female" (Sadker and Sadker 1986; Sadker, Sadker, and Hicks 1980).

The National Assessment further documented the educational inequities in mathematics and science for minority students, especially for African American and Hispanic children. Performance differences are evident for a large segment of minority students, beginning in elementary school and escalating over time. These students achievement scores fall significantly below the national average. Nationally, minority students take about one year less of high school mathematics and science courses than their peers. Home, school, and community are contributing factors to poor performance, which limits their pursuit of advanced study in high school. These students are not encouraged nor are they prepared to enroll in courses that will increase their competence and assist them in acquiring future technical positions. Most African Americans who pursue science and engineering programs do so at historically African American colleges and universities. Hispanics lack even this network of nurturing (Task Force on Women, Minorities and the Handicapped in Science and Technology 1988).

The rate of attrition for African American students in high school mathematics courses has reached crisis proportions and must be addressed. "There is little doubt that a failure to become mathematically literate will have disastrous consequences for these drop-outs as they enter the nation's labor pool. . . . Young Black people are threatened with exclusion from the worldwide technological revolution" (Williams 1985).

The research literature on educational progress in mathematics and science for physically handicapped students is almost nonexistent. Government reports show only a handful of physically handicapped students who are planning to enter high-tech fields of employment (United States Commission on Civil Rights 1983).

Unemployment rates for physically handicapped people are drastically higher than unemployment rates for the nonhandicapped. Current estimates suggest that between 50 and 75 percent of physically handicapped people are unemployed. The median family income of nonhandicapped individuals is nearly double that of the physically handicapped population. Moreover, African Americans and Hispanics are more likely than whites to have physical handicaps. Statistics also tell us that 12 percent of all adults have a limited physical handicap and an estimated 10 percent of all school-age children are physically handicapped (United States Commission on Civil Rights 1983).

Physically handicapped students enjoy as high achievement in the elementary grades as do females, but they rarely enter the employment pool in mathematics or science. With the aid of high technology and computers, physically handicapped persons would have few limitations in the workplace. Again, the double bind of being physically handicapped and female creates the most persistent barrier to advancement.

Underutilizing America's female, minority, and physically handicapped population wastes precious human resources. All students must learn to value mathematics and science and become confident in their mathematical and scientific ability. A better understanding of special populations and the factors related to their achievement in mathematics and science are required to improve the chances for female, minority, and physically handicapped students' academic and career success. Few intervention programs directly address females, minorities, and the physically handicapped in mathematics and science at the junior high and middle school level.

Project Overview

The project has been labeled TEAMS for its three-pronged teaming approach to combining the various programs, participants, and consultants to meet the equity needs in school districts (fig. 12.1). Conceptually, the most basic teaming component is the substantive blending of elements from cadres of trainees representing junior high and middle schools. Each cadre consists of seven members representing a central office administrator, a building-level administrator, a science teacher, a mathematics teacher, a counselor, a student, and a parent. Another component consists of consultants, advisors, and role models who assist in developing and implementing the training sessions.

The participants in Project TEAMS attend a total of forty hours of training in the TEAMS model, that is, one day a month for five months. One day of training is provided in each program with the exception of EQUALS, in which participants receive two days of training. Every participant receives a set of materials from each program. The materials provide a wealth of classroom activities in the areas of mathematics and science, and teachers are free to select activities that are appropriate to their class. Trainees participate in an additional eight hours of peer observation and performance-feedback coding

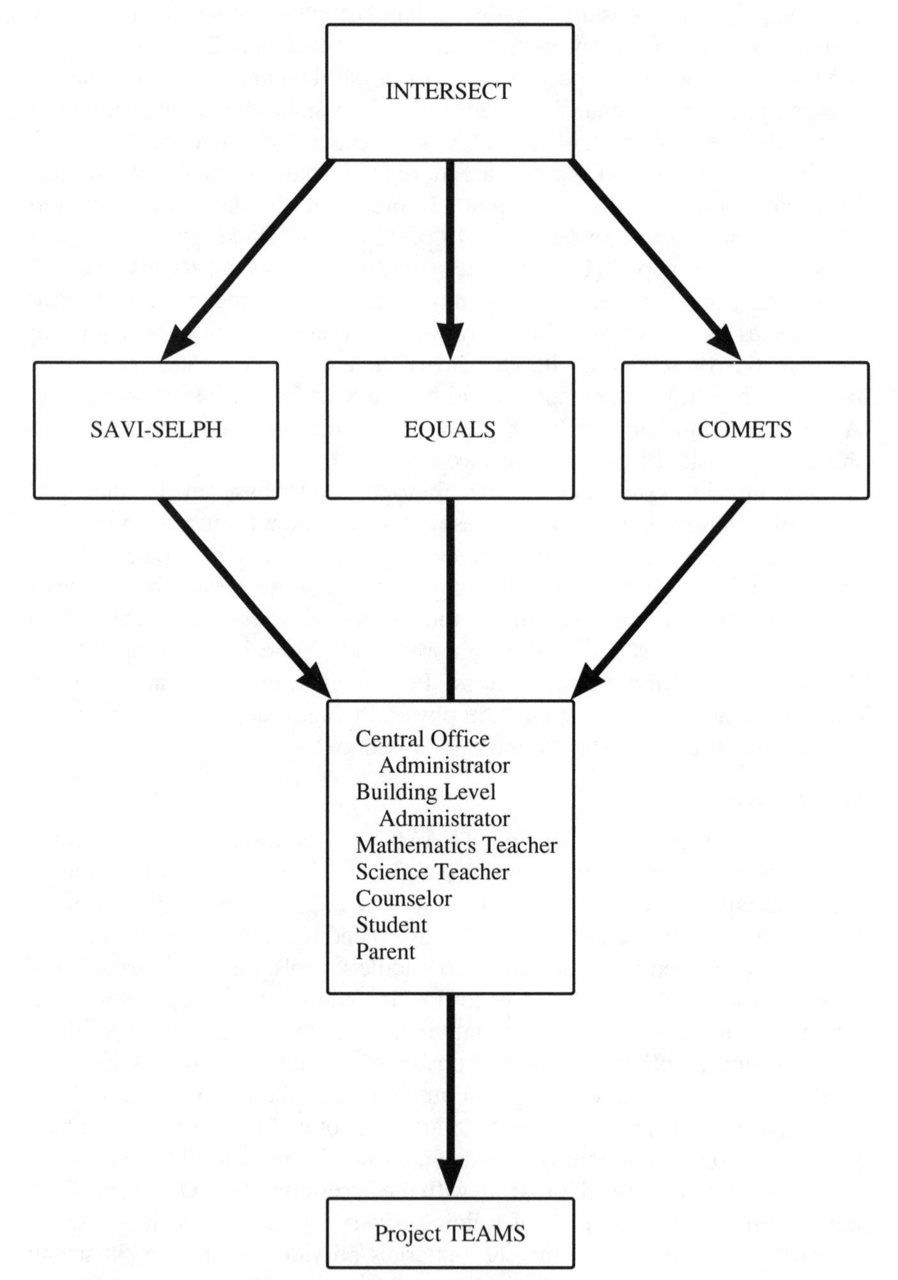

Fig. 12.1. Project TEAMS Overview

that focus on sex bias in student-teacher interactions. With the exception of SAVI-SELPH, national consultants present the training for their respective models. SAVI-SELPH training is provided by trained in-house staff. Sessions are limited to thirty-five participants for effective instruction and interaction.

The first teaming component involves blending elements from the model programs. Each program furnishes a unique component to the overall project. The purpose of INTERSECT (Interaction for Sex Equity in Classroom Teaching) is to assist teachers and counselors in achieving sex equity in their interactions with students. Participants are trained with diagnostic tools to analyze patterns of inequitable student treatment and are offered strategies to improve teaching and classroom organization. A peer-evaluation component gives performance feedback. EQUALS (Sex Equity Program in Mathematics, Technology and Career Education) provides activities that encourage the participation of students and adults, particularly females, in mathematics courses and promotes their interest and involvement in mathematics-based fields of study and work. The purpose of the COMETS (Career Oriented Models to Explore Topics in Science) program is to encourage early adolescents, particularly females, minorities, and the physically handicapped, to develop a broad knowledge of science concepts, career options, and community resource persons who work in science-related fields and a keen interest in the pursuit of science study and growth. SAVI-SELPH (Science Activities for the Visually Impaired and Science Enrichment for Learners with Physical Handicaps) provides a series of multisensory science enrichment activities for both handicapped and nonhandicapped students. The specialized equipment and procedures ensure full access to science learning for all students. The training in each program is designed to supplement the classroom mathematics and science curriculum.

The peer observation and performance-feedback-coding component is most significant to the practical application of the training techniques. Participants are trained to use the INTERSECT assessment instrument. The purpose of the instrument is to determine patterns of possible sex bias or inequity in the ways teachers and students interact with each other in classrooms. The instrument is divided into three sections. The first section requires the recording of descriptive data (e.g., grade level, subject area, a diagram of the class seating chart noting the race and sex of each student, and so on). The second section is the coding of teacher and student interactions. Although the coding requires practice, it describes the initiation of the interaction by the teacher or the student (identified by race and sex) and the form of the interaction (i.e., praise, acceptance, remediation, or criticism). The final section is a reporting of anecdotal descriptions of incidents, which yields rich data on behaviors and activities that may hinder or promote the achievement of sex equity in classroom interactions. The completed assessment instrument reveals patterns of student-teacher interaction in the observed classroom.

Following each training session, the project TEAMS participants develop and extend their skills by practicing equitable classroom interactions, implementing nonsexist behaviors and activities, and observing peers by using objective and specific information concerning teaching behaviors that promote or eliminate bias in the classroom. Participants with teaching assignments are released from classroom duties and provided with substitutes to enable them to attend all observation and coding sessions.

The second teaming component includes the cadres of trainees representing junior high and middle schools. Each participant represents an element needed to support the project. The central-office administrator and the building-level administrator are necessary to maintain district- and building-level support of the project. Without administrative support, projects tend to dissolve. Both the science teacher and the mathematics teacher are necessary to deliver instruction. The counselor plays a vital role in career orientation. Students and parents (not always related) provide the support and enthusiasm for the project among their respective groups. Every attempt is made to include members of the target population among the Project TEAMS participants. The final teaming component includes the consultants, the advisors, and the role models who assist in developing and implementing the training sessions. Role models furnish an exciting component to the project. Examples of role models might include pilots, medical researchers, engineers, plumbers, or actuaries, who provide lively discourse on their childhood, academic preparation, and professional experiences in achieving their nontraditional career roles. The role models often become an integral part of the teacher's classroom.

Project TEAMS attempts to achieve the five general goals developed for all students by the National Council of Teachers of Mathematics (NCTM 1989) but broadens the goals with the inclusion of science. These goals may be restated in the following manner:

1. Students learn to value mathematics and science.
2. Students become confident in their ability to do mathematics and science.
3. Students become mathematical and scientific problem solvers.
4. Students learn to communicate mathematically and scientifically.
5. Students learn to reason mathematically and scientifically.

The project was developed for junior high and middle school students and thus is designed to meet the needs of young adolescents. Kerewsky and Lefstein (1982) identified seven developmental needs of adolescents. Project TEAMS, as indicated by participants, developers, and consultants, meets the identified developmental needs of adolescents by promoting positive social interactions with adolescents and adults through parental involvement and role models yet provides structure and clear limits with meaningful participation particularly in the exploration of careers. Project TEAMS also promotes creative expression, physical activity, self-definition, and a feeling of competence and achievement for the participating adolescents. Adolescents are involved in hands-on activities that promote, for example, exploration and estimation. (Students may use estimation skills in an attempt to find the combined height of all students in the class, or they may explore creative problem solving with the challenge given to teams of students to build the highest tower using 8.5″ × 11″ paper, paper clips, scissors, masking tape, and a marking pen.) Through the use of such daily problems, adolescents learn the practical application of mathematics and science and realize that a lack of knowledge in these fields closes the door to many nontraditional careers.

Finally, Project TEAMS exhibits most of the characteristics identified by the American Association for the Advancement of Science (AAAS 1984) as those that produce successful mathematics and science intervention programs for underrepresented populations. These characteristics include—

1. a strong academic component in mathematics, science, and communications, focused on enrichment rather than remediation;
2. academic subjects taught by teachers who are highly competent in the subject matter and believe that students can learn the material;
3. a heavy emphasis on the applications of science and mathematics and on careers in these fields;
4. an integrative approach to teaching that incorporates all subject areas, hands-on opportunities, and computers;
5. multiyear involvement with students;
6. a strong director and a committed and stable staff who share program goals;
7. a stable long-term funding base with multiple funding sources;
8. the recruitment of participants from all relevant target populations;
9. a university, industry, school, and so on, cooperative program;
10. opportunities for in-school and out-of-school learning experiences;
11. parental involvement and development of a base of community support;
12. specific attention to removing educational inequalities related to gender and race;
13. the involvement of professionals and staff who look like the target population;
14. the development of peer support systems (through the involvement of a critical mass of any particular kind of student);

15. evaluation, long-term follow-up, and careful data collection;
16. "mainstreaming"—integration of program elements supportive of women and minorities into the institutional programs.

Project TEAMS displays the AAAS characteristics in a variety of ways. The selected programs incorporated into the project furnish enrichment rather than remediation activities, applications of science and mathematics to careers, an integrative approach using hands-on activities, and opportunities for in-school and out-of-school learning experiences. Participants and project staff demonstrate their competence and commitment to race and sex equity and the belief that all students can learn through their ongoing creative efforts. Role models, consultants, participants, and students are representative of the target population. Parents and students are recruited as members of each school cadre for their meaningful participation and support. University, school, and industry partnerships are formed through the role model connections. Data collection and evaluation are integral to Project TEAMS. The integration of Project TEAMS into the institutional program allows for the multiyear involvement of the students, which results in long-term gains as documented in the findings and anecdotal data section of this chapter. The stable long-term funding base is the most difficult characteristic to achieve, which is often the case in the funding of educational programs.

Obviously, a template of a program cannot be superimposed in a location. Local conditions must dictate the specific factors that will enhance change and effective learning. However, the components of Project TEAMS are flexible and replication is not difficult once the materials are available.

Findings and Anecdotal Data

The overall research question in the study was, "Does Project TEAMS training encourage females, minorities, and the physically handicapped in mathematics and science?" Although the findings from quantitative analyses were inconclusive, anecdotal data showed a variety of teacher and student comments about factors that continue to hinder females, minorities, and the physically handicapped in mathematics and science.

Qualitative data were collected through observations of, and interviews with, mathematics and science teachers who participated in the Project TEAMS training. The interview data were coded to define descriptors for content analysis, a standard technique for the organization of qualitative data.

Overall, the teachers were shocked at the inequitable interactions observed in their classrooms. An immediate response from one participant after the initial observation was, "If my interactions with students display this level of inequity, what is happening in the classrooms of fellow teachers who lack any knowledge of equitable student interactions?" Another respondent referred to the observations as an "eye-opener." As a department chair, this respondent used the technique in observing teachers and was shocked at what was observed in two young male teachers' classrooms, as indicated by the comment, "I thought they were great teachers until I observed the level of inequitable interactions in their classrooms." Now, every year, the respondent provides staff development on equitable classroom interactions. There were mixed attitudes as to the knowledge of teachers about equitable classroom interactions. One respondent replied, "Teachers are aware of race, sex, and handicapped equity but they are so overburdened—it is way down on the list of things to do." The two newer teacher respondents believed they had little or no preparation in equitable classroom interactions in their undergraduate training. One respondent replied, "As a new teacher you focus on student discipline, not equity." The other respondent stated, "I am a feminist. I know inequity exists but I don't know the statistics. Overall, I feel teachers are not knowledgeable of equitable classroom interactions."

The observed teachers often criticized or suggested remediation for the responses of females and minorities. White males were encouraged to respond, and their responses were accepted and praised. Praise for males was often in the form "You've done an excellent job in setting up the experiment," whereas females were praised for their neat papers. These responses reinforce sex stereotypes. The acceptance of responses to teacher-initiated questions was often more positive for males than for females and minorities as indicated by the enthusiasm in the teacher's voice. A remediation response indicates that the teacher does not accept the accuracy of a student's characteristic or behavior. "If you write with a sharper pencil, your paper will be neater" was an example of a remediation response that fails to comment on the intellectual quality of the work but comments on neatness, which is perceived as a female quality. Criticism indicates a negative teacher evaluation and strong disapproval. "I won't accept this paper; the answers are inaccurate and it is sloppy" indicated a criticism of both intellectual quality and appearance. The participants' level of awareness of inequitable student interactions and their classroom actions exhibited marked improvement during the period of intervention and continued throughout the school year.

Additionally, seat assignments in some classes were found to support inequitable interactions between students and teachers. This usually occurred when students worked in pairs or groups, such as in lab settings, where males and females may sit at separate tables and the teacher may give more attention to the group of male students. This problem was remedied by all teachers on identification. Reassigning students allowed for more equitable interaction.

Interesting factors emerged when teachers were questioned about internal and external factors that hinder females, minorities, and the physically handicapped in mathematics and science. None of the respondents indicated an internal, or school-related, factor specifically at the junior high or middle school level. All the respondents indicated internal factors at the elementary school level, and one respondent also stated that the secondary school was the problem.

In response to the internal factors, the respondents made the following statements:

"Elementary teachers feel incompetent in math and science skills."

"Elementary schools are doing a terrible job at math and science skills. People who were good at subjects moved into junior high/middle schools."

"We must start early to change attitudes at the elementary level."

"We need to start programs at a younger age, maybe grades four or five."

"We need to start teaching science at the elementary level. Elementary teachers don't feel competent and confident."

Overwhelmingly, the external factor indicated was the parents, in particular, the mother.

It is interesting to note that three of the respondents above were elementary school teachers who moved into junior high or middle school teaching. Two respondents indicated that elementary school teachers should be encouraged to attend conferences such as those sponsored by the National Council of Teachers of Mathematics (NCTM) or the National Science Teachers Association (NSTA). One respondent has started a staff development program in science during the summer for teachers from the feeder elementary schools, but the respondent states "those who already do a good job come."

Parents were seen as reinforcing the stereotypes that mathematics and science are white male domains. As one respondent indicated, "Girls get the message from mom. It's OK if you are not good in math and science, I'm not good in it, either." Another respondent stated, "Some parents feel a girl can be good in math or science but never as good as her brother." "Parental attitudes are poor," stated a respondent, "probably more so for mothers. Mothers lack confidence, even though they manage homes. Generation to generation—it's like the dark ages. I could never read, so it is OK for you not to read." To highlight the lack of mothers' support, one respondent stated, "On family math nights, more fathers than mothers will come." On a positive note, one respondent stated, "In parent conferences when discussing their child's potential, I can talk about the research from Project TEAMS and parents listened." Many of the interviewed teachers organized and conducted parent activities to inform parents of their role in promoting their children in mathematics and science and encouraged parent participation in activities that fostered involvement of their children in mathematics and science. Some activities were conducted in a school district where a teacher job action prohibited after school activities due to the failure to reach a settlement on the teacher contract. Two teachers in one school involved other teachers and initiated a family night at school in all subjects. They provided child care during the parent sessions. These two energetic and creative teachers organized and gained school board support for a family mathematics lab, which was opened one night a week during the school year. Another teacher initiated a summer science camp, which included many field trip experiences for students to visit nontraditional workers in mathematics and science-related careers. In their classrooms, these teachers directed intervention toward inequity issues that before Project TEAMS they may not have recognized. These teacher participants were extremely motivated to be effective teachers and were committed to equity in their classrooms.

Additional qualitative evidence emerged during site visitations and informal interviews. In the school that previously had the lowest mathematics and science achievement test scores for middle schools in the district, girls' achievement test scores in science increased by 12 points following Project TEAMS. The participants also reported an increase in female and minority participation in their annual science fairs. In one school, all sixth graders are required to participate in the science fair. Science teachers spend time throughout the semester helping the students develop and organize science projects. For the first time, female, minority, and handicapped role models were recruited as judges for the science fair. Furthermore, all the schools have indicated a small yet significant increase of female and minority students in science and mathematics classes for gifted students. Several of the middle schools in this study feed into one high school that recently increased the number of physics classes from two to eight sections.

Project TEAMS's participants have enthusiastically implemented the curriculum models in their classrooms, with nontrained teachers clamoring to borrow the materials. The peer-observation model has continued to provide an ongoing vehicle for eliminating bias in teacher-student interactions. Through the building "team" cadre, schools have also made strides in promoting computer equity, career awareness, hands-on participation in science activities for both males and females, and increased exposure to nontraditional female, minority, and physically handicapped role models. One school has implemented a program where a role model is engaged monthly as a presenter at a grade-level assembly. The students have displayed a heightened interest in locating role models in careers of interest to them. Some students have undertaken the task of communicating on a regular basis with specific role models to gain additional insight into particular careers.

Conclusion

Most universities require courses in advanced mathematics and science for many majors. Students who avoid mathematics and science courses close off a wide range of future career options. Avoiding mathematics and science courses does not occur until high school, where such courses are no longer mandatory. Therefore, it is imperative that females, minorities, and the physically handicapped are convinced during the junior high or middle school years, when attitudes are forming, that mathematics and science courses are enjoyable, appropriate for all students to pursue, and useful in attaining career goals. Because mathematics is a sequential subject and many sciences require a mathematics background, students must be encouraged to study these subjects early and to continue those courses of study throughout their school years. Students, particularly minorities, who become locked in general or remedial classes are usually unable to participate in advanced academic coursework (Beane 1985).

Kahle (1985) studied the retention rates of girls in science through case studies of secondary school teachers. She states, "There are no simple explanations for her (teachers') success." Throughout her case studies, talent, commitment, rapport with students, and ability to motivate were the factors associated with a teacher's encouragement of females in mathematics and science. These qualities were continually exhibited in the Project TEAMS participants.

Racism, sexism, and prejudice against the handicapped have limited many career opportunities. The leadership of the United States cannot be maintained unless females, minorities, and the physically handicapped are viewed as essential contributors to national strength in mathematics, engineering, and science. The key to this strength is the education pipeline from kindergarten to graduate school. Local intervention programs such as Project TEAMS must be developed, disseminated, and adjusted to local conditions with the goal that females, minorities, and the physically handicapped will be adequately represented in all fields of study.

References

American Association for the Advancement of Science. *Equity and Excellence: Compatible Goals. An Assessment of Programs That Facilitate Increased Access and Achievement of Females and Minorities in K–12 Mathematics and Science Education.* Washington, D.C.: Office of Opportunities and Science, 1984.

Beane, DeAnna B. *Mathematics and Science: Critical Filters for the Future of Minority Students.* Washington, D.C.: American University, Mid-Atlantic Center for Race Equity, 1985.

Fox, L. H. "The Effects of Sex-Role Socialization on Mathematics Participation and Achievement." In *Women and Mathematics: Research Perspectives for Change,* NIE Papers in Education and Work, No. 8, Washington, D.C.: National Institute of Education, 1977.

Horner, Matina. "Sex Differences in Achievement Motivation." Ph.D. diss., University of Michigan, 1968.

Kahle, Jane B. *The Disadvantaged Majority: Science Education for Women.* Princeton, N.J.: American Educational Testing Service, 1983.

________. "Retention of Girls in Science: Case Studies of Secondary Teachers." In *Women in Science: A Report from the Field*, edited by Jane B. Kahle, pp. 49–76. Philadelphia: The Falmer Press, 1985.

Kerewsky, W., and Leah M. Lefstein. *Young Adolescents and Their Communities: A Shared Responsibility.* Chapel Hill, N.C.: University of North Carolina, Center for Early Adolescence, 1982.

Mandelbaum, Dorothy R. *Work, Marriage and Motherhood.* New York: Praeger Publishers, 1981.

National Assessment of Educational Progress. *Third National Mathematics Assessment: Results, Trends and Issues.* Princeton, N.J.: Educational Testing Service, 1983.

National Council of Teachers of Mathematics. *Curriculum and Evaluation Standards for School Mathematics.* Reston, Va.: the Council, 1989.

National Research Council, Office of Scientific and Engineering Personnel. *Minorities: Their Underrepresentation and Career Differentials in Science and Engineering.* Washington, D.C.: National Academy Press, 1987.

Sadker, Myra, and David Sadker. "Sexism in the Classroom: From Grade School to Graduate School." *Phi Delta Kappan* 67 (March 1986): 512–15.

Sadker, Myra, David Sadker, and T. Hicks. "Sex-Equity in Teacher Preparation." *Journal of Teacher Education* 31 (November 1980): 4–5.

Task Force on Women, Minorities and the Handicapped in Science and Technology. *Changing America: The New Face of Science and Engineering.* Interim report. Washington, D.C.: U.S. Government Printing Office, 1988.

United States Commission on Civil Rights. *Accommodating the Spectrum of Individual Abilities.* Clearinghouse publication 81. Washington, D.C.: U.S. Government Printing Office, 1983.

Williams, James D., ed. *The State of Black America—1985.* New York: National Urban League, 1985.

Sources of Program Information

COMETS (Career Oriented Modules to Explore Topics in Science). The model is available through the National Science Teachers Association, Washington, D.C.

EQUALS (Sex Equity Program in Mathematics, Technology and Career Education). The model is available through the Mathematics/Science Network, Berkeley, Calif.

INTERSECT (Interactions for Sex Equity in Classroom Teaching). The model is available through The Network, Inc., Andover, Mass.

SAVI-SELPH (Science Activities for the Visually Impaired and Science Enrichment for Learners with Physical Handicaps). The model is available through the Lawrence Hall of Science, University of California, Berkeley, CA.

13

Helping Inner-City Girls Succeed: The METRO Achievement Program

Denisse R. Thompson
Natalie Jakucyn

To reach all students with mathematics, it is essential that all students have access to the support facilities that will enable them to succeed. Gifted students generally have such access, since many programs exist that are designed to foster the special talents that such students have. Remedial students also have access to special programs designed to help them overcome learning problems or deficiencies in their background. However, *average* students often fail to receive additional support and thus fall through the cracks in the educational system.

The general lack of additional educational support for *average* students becomes an even more serious problem in the inner city, where such students are generally of a lower socioeconomic status, are members of a racial or ethnic minority, and attend schools in financially troubled educational systems. Although there is a lack of research on the interaction of these factors, research on the effects of social class or ethnicity alone suggests that inner-city students, both male and female, face a number of obstacles in obtaining a quality education that keeps educational and career options open. Baron, Tom, and Cooper (1985) indicate that the first impressions of a teacher may be related to the gender, race, and socioeconomic class of the student. And "the race or class of a particular student may cue the teacher to apply the generalized expectations, therefore making it difficult for the teacher to develop *specific* expectations tailored to individual students" (p. 251). In general, teacher expectations and achievement standards are lower for African American students and lower-class students than they are for white and middle-class students (Baron, Tom, and Cooper 1985; Scott-Jones and Clark 1986).

In addition to the overall lower expectations, many African American and Hispanic children, particularly those in the inner city, attend schools that are predominantly minority (Arias 1986; Chicago *Tribune* 1988; McBay 1990). Such schools are often overcrowded and lacking in modern facilities and current materials (Matthews 1984; McBay 1990). Arias (1986) found that the attrition rate in Chicago, as well as in other urban centers such as New York and Los Angeles, is worst at schools "with high concentrations of low-income students, with low reading scores and with over 70 percent minority enrollment" (p. 40). The result is a spirit of hopelessness that pervades such schools.

Also of concern is the nature of classroom instruction as it relates to the cultural background of minority students. School culture often expects and emphasizes behavior patterns that are based on middle-class white values that may conflict with the culture in which students live (Silverstein and Krate 1975; Slaughter and Epps 1987). Students from various racial and ethnic backgrounds need to learn that they can be successful in school without giving up their cultural identity (Ogbu 1985).

The typical gender bias that females face in mathematics and science is another obstacle added to those already experienced by inner-city children. Not only do females take fewer advanced mathematics courses than males (Campbell 1986; Meyer 1989), but their expectations to succeed decrease as they pass through junior high school. In fact, for girls, decreases in the confidence of their ability and their expectation to succeed "were found to *precede,* rather than follow, the decline in math achievement by girls" (Campbell 1986, p. 518).

In addition to lower expectations for females and minorities, there are hindrances to learning that are particular to the mathematics classroom. Instruction in mathematics is frequently characterized by teacher lecture, independent student seat work, and an emphasis on tests. In contrast to this pattern, a classroom that accounts for differences and preferences in learning, such as cooperative learning, may be especially beneficial to minority students (Syron 1987; Stiff 1990). A classroom that emphasizes the opportunities and jobs that mathematics study supports provides minority students with essential information needed to maintain a motivation to learn (Syron 1987; MacCorquodale 1988).

The METRO Achievement Program

In response to these problems, a group of concerned educators founded the METRO Achievement Program in Chicago in 1985. The METRO program primarily serves African American and Hispanic girls from the west side of Chicago. Since 1962 there has been a program in this area for boys in grades 7 through 9. Although originally started as a tutoring, sports, and trades program, it expanded in 1972 to emphasize academic achievement and character development. When concern was expressed that girls in these neighborhoods needed similar opportunities, it was decided to form a separate program for girls rather than to make the program coeducational. With adolescents, it was felt that more could be accomplished in single-sex programs.

The METRO Achievement Program is designed as a supplementary educational program to help average girls entering the seventh and eighth grades "who are receptive to challenge and seek to develop their full potential" (METRO 1987). The emphasis on the seventh and eighth grades reflects the pivotal role that these grades play in the educational choices of students, particularly in relation to the study of mathematics and science. Nicholson (1986) writes that, for females, "participation in math and science—in education and employment—remains disproportionately low. For girls from minority and low-income backgrounds [characteristics of inner-city girls] the problem is especially acute" (p. 1). Further, she indicates that "the junior high/middle school years are the critical period when patterns for dropping out of math and science become entrenched" (p. 2). Brush (1979) reports that students have ambitious ideas about continued mathematics study, with boys planning to take more mathematics than do girls, even by the seventh grade. Hence, it appears that socialization influences about appropriate academic paths for females are already being internalized, even at this early stage in the educational process. Nelson, in a report for the Girls Clubs of America, sheds additional light on this problem by indicating that the time period from grades 5 or 6 to grade 8 is a time "when girls are most vulnerable to peer values about what it means to be competent, successful, and female. Classroom and out-of-school experiences which consistently communicate stereotyped or limited expectations or are sparse in stimulating examples and encouragement will undercut girls' motivation to learn and will undercut the development of their full potential as productive and self-actualizing adults" (Nelson 1987, pp. 3–4).

As mentioned previously, METRO focuses on average girls entering the seventh and eighth grades. Its program is designed to give girls the needed support and the ammunition, both academic and social, that they need to avoid the limitations and stereotypes referred to by Nicholson, Nelson, and Brush. Indeed, the program has three broad goals designed to provide the type of academic stimulation with high expectations that is often missing from many inner-city school settings. These goals are to *encourage academic achievement, to aid personal development, and to instill a spirit of community service.* Through these goals, METRO works to help girls "qualify for entrance into the best high schools [in the city] and eventually, colleges" (METRO 1987).

The METRO Achievement Program is funded by private sources (foundations and corporations) and is without affiliation with any educational institution.

To describe the METRO program, this chapter is organized into four parts: a description of the general characteristics of the program; a description of the specific features of the mathematics strand; a description of a typical day at the program; and a discussion of some evaluation results.

General Program Characteristics

The METRO program runs five days a week for five weeks in the summer. (There is also an aspect of the program that exists throughout the school year for girls in grades 4–12. That aspect of the program is not discussed in this chapter.) About eighty junior high school girls, half entering the seventh grade in the fall and half entering the eighth grade, participate in academic classes (mathematics, science, and communication skills), a character development class, a fine arts class, team sports, and personal counseling sessions. There is also a weekly excursion to a place of interest in the Chicago metropolitan area. The faculty consists of four permanent administrators (two of whom teach a class), five teachers, and seven high school girls who act as tutors. In addition, speakers come to the program and furnish information about various careers; occasionally these speakers serve as adult mentors for girls who desire to enter the discussed professions. The career speakers are important in helping girls realize the academic requirements that are necessary for success in many fields.

The academic classes are designed to improve skills that may enable the girls to enroll in an academic track in junior high and high school, a decision that is at the heart of keeping career and educational options open. There is also an attempt to improve attitudes toward school and toward school subjects. It is hoped that this academic work will translate to improved grades during the following academic year by building a link between attitude/motivation and achievement. Eccles and Wigfield (1985) have characterized this link as a feedback system in that "positive motivation facilitates achievement, which, in turn, facilitates continued positive motivation" (p. 186).

Although this chapter will focus primarily on the mathematics portion of the program and its effects, a brief description of all the classes follows to give an overall picture of the program in which mathematics is embedded.

Mathematics: The curriculum is applications-based with a strong technology component. Mathematics is the only class that uses a textbook.

Science: The class emphasizes laboratory work oriented toward the natural and physical sciences.

Communication skills: The class includes writing and grammar and emphasizes public speaking.

Character development: The curriculum has a religious base with frank discussions about sexual morality, relationships with friends and family, and relationships with God.

Fine arts: This class provides an outlet for girls to express themselves through artistic media.

Team sports: This is an active period in which girls learn the rules for, and participate in, various sports.

The personal development aspect of the program is handled through one-on-one counseling to help girls develop "clear convictions, loyalty, thoughtfulness, consideration of others, and a readiness to be of service" (METRO 1987). Counseling occurs concerning sexual morality, drugs and alcohol, relationships with friends and family, interests, and academic plans. Discussion about the first two issues is designed to make program participants aware of the devastating consequences that result from early pregnancy and drug or alcohol abuse.

Recruitment, Admission, and Participant Population

METRO recruits participants for the program through spring visits to most of the public and parochial schools within a three-mile radius of Chicago's downtown area. School visits depend on the willingness of school personnel to allow the presentation. After the presentation, girls complete a brief survey indicating whether they would be interested in attending METRO. Program personnel then attempt to contact by phone the families of interested girls to set up interviews with the girls and their parents.

During the participant interview, girls are questioned about their attitude toward school and about their reasons for wanting to attend METRO to determine if those reasons are in agreement with METRO's goals. In general, METRO tries to identify that "good kid who just needs a chance."

Selection is based on participant and parent interviews, together with school grade and attendance records. Program personnel look at consistency of grades, marks relating to effort and behavior, the number of days absent or tardy, and any teacher comments. Scores from a standardized test administered during the school year are also used. Selected girls need to have a composite score at least at the fortieth percentile. In addition, the component verbal and quantitative scores are examined because METRO usually does not accept girls low in both areas. The rationale for this decision is that a high score in one area gives METRO a basis for development. Although a fee is charged for the program, scholarships ensure that no girl is denied an opportunity to attend the program solely because of financial need.

The final composition of the participant population is a critical aspect of the METRO program that, together with the curriculum, is responsible for much of the program's success. Participants come from a variety of ethnic and racial backgrounds (about 30 percent African American, 60 percent Hispanic, and 10 percent Caucasian, Asian, or Native American). This mix affords an opportunity for multicultural exchange as well as a chance for girls to interact in a close environment with individuals from outside their own ethnic or racial group. For the first time, many of the girls make a friend who is from another race.

The fact that the participant population is entirely female is another critical aspect of the program. (The staff is also entirely female.) Especially during these adolescent years in school, boys tend to act as social distractors or as people against whom to compete. In the METRO program, girls are forced to compete academically with each other and to take leadership roles. There are no males to hinder academic performance in mathematics or science.

A final characteristic of the participant population is that girls come from nearly thirty different schools. Hence, very few of them enter the program knowing any other participants. This allows each girl to start the program without academic or social preconceptions of other participants.

Parental Involvement

Parental involvement is another important feature of the METRO program. This involvement begins with a parent interview during the selection process. Parents are informed of the academic expectations of the program, such as a new textbook (in mathematics), the use of a calculator, required homework, and regular tests. Paren'ts make a personal, verbal commitment indicating a willingness to help their daughter by encouraging her, monitoring her academic and personal progress, and giving her academic assistance when possible. They also commit to seeing that their daughter attends the summer session daily and on time. If their daughter is irresponsible twice in any area (failure to complete homework or disciplinary problems), the parents are notified.

During each summer session, a Saturday workshop on *parenting* is offered. METRO helps parents to realize that they are primarily and principally responsible for the education of their children. This means that "in the family it is a question of educating what is most natural—the inner core of the person" (Isaacs 1984, p. 7). By analyzing case studies, parents learn how to handle various family situations—essentially, how to be a better parent. (Translators are available for those parents who do not speak English.) These workshops are given by model parents of METRO students or by other adults who have taken seminars on family enrichment. Usually, about one-third of the families are represented at these workshops.

The Mathematics Strand

To compete in today's world, individuals must be mathematically literate. For this reason, the METRO Achievement Program has incorporated mathematics classes as a standard part of its curriculum. METRO tries to instill into its girls a strong desire and capability for academic success and self-esteem. Thus, it is expected that every girl will succeed in mathematics.

Well before the *Curriculum and Evaluation Standards for School Mathematics* (National Council of Teachers of Mathematics 1989) was released, METRO shared key underlying assumptions with that document:

- No one "will be denied access to the study of mathematics . . . because of a lack of computational facility."
- There will be a "core [mathematics] curriculum differentiated by the depth and breadth of the treatment of topics. . . . "
- Mathematics will be taught in the context of applications.
- Mathematics topics will be integrated.
- Appropriate technology will be available to all. (pp. 124–25)

These assumptions necessitated a mathematics curriculum that comprised more than the basic skills. The textbook *Transition Mathematics* (Usiskin et al. 1986), developed by the University of Chicago School Mathematics Project, was chosen because it incorporates technology and real-life applications as well as other features deemed essential to the development of independent learners capable of success in mathematics. METRO's staff had preferred not to adopt a textbook to avoid having participants think of METRO as summer school. However, writing a new curriculum was deemed to be too time-consuming.

Mathematics Curriculum

The content of the mathematics program is that found within the first five chapters of *Transition Mathematics,* specifically chapters 1 and 2 for seventh graders and chapters 4 and 5 for eighth graders. This results in a seventh-grade focus on applied arithmetic (decimals, fractions, percent, scientific notation) and an eighth-grade focus on algebra (patterning, variables, integers, open sentences), pregeometry (angles, polygons, perimeter, area), and a continuation of applied arithmetic. Several curriculum features, central to the materials developed by the University of Chicago School Mathematics Project, are common to the program at both grade levels.

First, the use of a *scientific calculator* is assumed, and the program provides a scientific calculator for each girl. The calculator serves as a great equalizer because it places all girls on an equal footing in terms of computational ability.

Second, *applications* are not just "added on" at the end of a lesson but are used to develop concepts and practice skills so that students need not ask "When will I ever use this?" For instance, in applied arithmetic, a discussion of estimation and the importance of rounding begins with five reasons for estimating and some corresponding situations for which an estimate may actually be preferable.

Third, applications lead to the *integration* of topics. Besides connecting mathematics to other such disciplines as social studies and science, applications can tie together branches of mathematics—such as geometry, algebra, and arithmetic—to provide "the big picture" of how and why the branches fit together. For instance, when situations involving addition are discussed, applications of distance can be extended to finding the perimeter of geometric shapes and then connected to algebra by finding unknown dimensions of a figure with a given perimeter.

Fourth, because classes are only thirty-five minutes long and meet only four times a week, the *pace* of each course is rather fast. A fast pace is possible because a *distributed review* section is included in each lesson to assist participants in retaining the content. In the program, no big chunk of time allows for practice. Rather, drill is broken into bite-sized pieces repeated throughout the entire session.

Fifth, because METRO is a supplementary program, it strives to "wean" participants from depending on the teacher or the program so that they can work independently in their formal school environment. Every participant is expected, and taught how, to *read* a mathematics textbook, no matter what her reading level. This enables the participant to initiate the learning of mathematics. "To succeed, you must read!" is the slogan of the mathematics classes.

Sixth, to motivate participants and to determine each participant's progress, various *classroom evaluation* measures are used. Daily homework is assigned and graded. Quizzes and tests are administered at least weekly. Strategy games are used to measure analytical processes and spatial visualization and to foster a competitive spirit. A mathematics award (a scientific calculator) is presented during METRO's graduation ceremony to the best student in each mathematics class.

Classroom Physical Environment

Although METRO is an academic intervention program, it tries as much as possible not to have an institutional atmosphere. Generally, the program meets in a shopping mall or in a converted warehouse. Only as a last resort does the program meet in a school.

In the classrooms, long rectangular tables are used instead of desks to encourage communication and cooperation among participants. In addition, the walls of the mathematics classroom are decorated with posters that apply mathematics to three areas: to professions, such as engineering and interior design; to advanced mathematics, such as stellated polyhedra or the logarithmic spiral in nature; and to artistic and visual works, such as Escher prints.

The classroom also contains a variety of games and mathematical toys to enhance analytical processes and spatial visualization and to encourage a certain level of competitiveness. These games are available for participants' use before and after the class, at the beginning and end of the day, and after early completion of a test or quiz. Some of the available games are checkers, chess, Mastermind, Othello, SOMA Cube, visual-thinking cards (cards containing mathematics that must be done visually), and Cuisenaire rods.

Staffing Criteria

Good teachers are vital to the success of the program. Professional mathematics teachers staff the mathematics classes. Only local area teachers are considered for staff positions because they understand the local school system and the high school possibilities that exist for the girls. In addition to possessing a strong content knowledge of mathematics, preferably with an undergraduate degree in mathematics or mathematics education, teachers must possess an educational philosophy that coincides with METRO's aims.

It is crucial that teachers have high expectations of participants and the conviction that *all* the girls can and should learn a good core of mathematics. That is, teachers must teach appropriate attitudes toward mathematics as well as the mathematics itself. The importance of teaching attitudes was eloquently expressed by Charles Merideth (1990) at the National Convocation for Making Mathematics Work for Minorities:

> [O]ur success with subject matter is determined by our attitude and the attitude of those we are teaching. Attitudes have their own distinctive aromas; students can smell them. . . . They know when we care. They know when we believe in them, they know when we don't believe in them and, in fact, they take their attitudes from our attitudes. (p. 61)

Hence, at METRO, the teachers' attitudes toward learning mathematics must be contagious.

In addition to helping girls develop a positive attitude toward mathematics, teachers must understand that educating a child in mathematics (as in any other subject area) necessitates educating the whole person, that is, building the character of each girl. Jacques Maritain (1943), a noted French philosopher, made the following statement about education:

> [The aim of education] . . . . is to guide man in the evolving dynamism through which he shapes himself as a human person—armed with knowledge, strength of judgement, and moral virtues—while at the same time conveying to him the spiritual heritage of the nation and the civilization in which he is involved, and preserving in this way the century-old achievements of generations. The utilitarian aspect of education—which enables the youth to get a job and make a living—must surely not be disregarded, for the children of man are not made for aristocratic leisure. But this practical aim is best provided by the general human capacities developed. And the ulterior specialized training which may be required must never imperil the essential aim of education. (p. 10)

Thus, besides teaching mathematics, teachers must work to foster such values as honesty, responsibility, generosity, and perseverance.

Teachers also need to see themselves as facilitators of learning and not simply as disseminators of knowledge. Each of them must truly be an "educator"—one who helps to form human persons, going well beyond the transmission of knowledge. This philosophy of the teacher's role is important in achieving METRO's goal of providing participants with the ability to initiate learning and to succeed when they return to their normal school environment. Consequently, teachers must maintain a friendly and cooperative, yet disciplined, atmosphere in the classroom.

Finally, teachers must be available to assist participants outside the normal class time—before and after the class, at the beginning and end of the day, during the weekly excursions, and at any other available time during the day when a girl needs assistance.

To ensure that teachers implement METRO's philosophy of education, several training sessions are provided. Besides the initial conversation during each teacher's personal interview, a one-day in-service session is conducted before the program begins. Practical guidelines are presented and discussed by the staff. There is a follow-up with daily staff meetings during the course of the summer program and periodic classroom observations by METRO's director. The daily staff meetings afford an opportunity for the staff to communicate about any problems that may be brewing and to discuss ways of resolving such problems before they become major obstacles.

Ability Grouping without Tracking

To foster equity in mathematics, METRO does not advocate tracking but it does promote ability grouping. Since tracking involves ability grouping with differentiated curricula, it violates the program's philosophy that all girls can learn a good core mathematics curriculum at their own pace and to different depths. Therefore, by using the same mathematics textbook and the same core content, which are rich enough to vary the depth of coverage, METRO can divide the classes by ability in a manner unbeknownst to the participants. At each grade the girls are split into two groups. The cutoff, though dependent on that summer's population, is determined by four factors: mathematics grade in school; score on a standardized mathematics test; participant writing sample and interview during the selection process; and teacher recommendation.

Within the groups, individual differences and deficiencies are met primarily by high school girls who act as tutors within the mathematics class. Many of these high school tutors are graduates of the METRO program and, thus, are believers in the influence the program can have on academic opportunities. These high school tutors not only provide immediate assistance but also act as role models for the participants in the program. They also lower the teacher-participant ratio so that a measure of personal attention is possible within the mathematics class.

A Typical Day at METRO

A typical day at METRO is packed with a variety of activities throughout the range of classes. As the schedule of classes (see fig. 13.1) indicates, there is no "fluff" or wasted time at the program.

Below is an abbreviated look at the activities in the eighth-grade classes on one day in the program, as reported in the evaluation report of the 1988 summer program (Thompson 1989).

10:00–10:54 Fine Arts: Girls are working on a project involving cans or bottles. Pieces of masking tape are torn into very small pieces and then layered on the outside of the container. When the container is completely covered, girls polish the tape with shoe polish. The texture of the masking tape picks up the polish to create an interesting holder for different objects.

On completing this project, girls begin to work on planning their goal banner. The goal banner is a major project that should reflect the highest aspirations of the girls. After mapping out the banner on newsprint, the girls will transfer designs to muslin.

11:04–11:58 Sports: Soccer is the sport of the day. Girls practice dribbling in a race format with girls in two lines. After practicing, girls are split into two teams for an actual game. Two high school assistants serve as team captains and make the choices for the team.

12:10–12:27 Lunch: Boxed lunches are provided by the city of Chicago. Most girls do not think too highly of the lunch provided.

12:30–1:06 Character Development: Class begins with a recap of the previous day's discussion on modesty. A discussion ensues about bringing modesty into relationships and the ways in which it exists in day-to-day living. Modesty can be acquired through human tools, such as dressing properly and choosing to associate with the proper people, as well as through spiritual tools, such as prayer. A quiz about issues discussed on the previous day ends the class period.

1:08–1:44 Science: Girls spend the class period making mobiles from straw, string, and cutout construction paper figures. The center straw is anchored in modeling dough or some type of clay. Other straws are used as cross beams. Girls work to position hanging paper pieces from the two ends of the horizontally placed straw so that the figures balance the straw.

Throughout, there is a discussion of possible uses of mobiles for party decorations. Girls work quite enthusiastically on the project and help classmates as necessary.

1:46–2:22 Communication Skills: Class begins by reviewing the previous day's activity, which involved creating sentences using specific words. The

	7A		7B		8A		8B
10:00–10:35	COMMUNICATION SKILLS		MATHEMATICS	10:00–10:10	GO TO PARK	10:00–10:54	FINE ARTS
10:37–11:12	MATHEMATICS		CHARACTER DEVELOPMENT	10:10–11:04	SPORTS	10:54–11:04	GO TO PARK
11:14–11:49	CHARACTER DEVELOPMENT		SCIENCE	11:04–11:14	GO TO METRO	11:04–11:58	SPORTS
11:51–12:27	SCIENCE		COMMUNICATION SKILLS	11:14–12:08	FINE ARTS	11:58–12:08	GO TO METRO
12:30–12:47	LUNCH		LUNCH	12:10–12:27	LUNCH	12:10–12:27	LUNCH
12:50–1:45	FINE ARTS	12:50–1:00	GO TO PARK	12:30–1:06	SCIENCE		CHARACTER DEVELOPMENT
		1:00–1:55	SPORTS	1:08–1:44	CHARACTER DEVELOPMENT		SCIENCE
1:45–1:55	GO TO PARK			1:46–2:22	MATHEMATICS		COMMUNICATION SKILLS
1:55–2:50	SPORTS	1:55–2:05	GO TO METRO	2:24–3:00	COMMUNICATION SKILLS		MATHEMATICS
2:50–3:00	GO TO METRO	2:05–3:00	FINE ARTS				

(35 min. classes; 2 min. changes; 17 min. lunch; 55 min. arts/sports; 10 min. drives)

(36 min. classes; 2 min. changes; 17 min. lunch; 54 min. arts/sports; 10 min. drives)

Fig. 13.1. METRO Achievement Schedule—summer 1988

focus of today's class is to write a news story. The news story created by the class is a report of speeches that had been presented by the girls at an earlier class. Some of the topics for the speeches, as reported in the news story, were "Saying No to Drugs," "Stopping Abortion," "Staying in School," and "Learning to Read."

Throughout the creation of the news story, the teacher gives lots of positive reinforcement and encourages interaction between teacher and students as well as among students.

2:24–3:00 Mathematics: Students are working on lesson 4-3 of *Transition Mathematics,* a lesson on translating algebraic expressions. Girls check answers with an answer key on the overhead projector while the teacher walks around the room to determine which problems created difficulties for the girls. These become the problems for class discussion. As the teacher circulates, she and the high school student assistants provide individual attention to the girls.

Evaluation

From the program's inception in 1985 until 1988, evidence of the program's success was essentially anecdotal. For instance, one parent marveled that a daughter who was generally poor at mathematics managed to excel and receive the mathematics award at the conclusion of the program. One fall, a teacher of one of the program's participants inquired about the change that had occurred in one of his students. The girl had changed from a student uninterested in academics to a student determined to succeed. In another instance, four of twenty eighth graders who had participated in METRO passed tests at their respective schools to enter an eighth-grade honors algebra class; these girls would otherwise have been in a nonhonors program. Additionally, METRO participants frequently commented on the personal attention and support that they received from the program staff, noting that for the first time, someone really paid attention to them.

For the summer 1988 program, a decision was made to attempt to collect information in a systematic manner and quantify it in some way. With that goal in mind, an evaluation was designed to obtain information on the program's effectiveness from a variety of sources: from the girls, from their parents, and from a few METRO graduates now in high school. Only a fraction of the data can be discussed here; more detailed information about the results (such as attitude measures by racial and ethnic group, self-esteem measures, and comments about nonmathematics classes) can be found in Thompson (1989).

Results from Participants

At the end of the program, participants were asked to respond to an open-ended questionnaire that solicited comments about the program in general and about specific classes. Sixty girls responded to the questionnaire, sometimes with more than one response for each question. In response to "If someone were to ask you whether or not they should attend METRO, what would you tell them? Why?" almost all the girls responded positively. Only four girls responded negatively. Girls indicated that METRO was fun and a new experience and that it was a place to meet new friends. Slightly more than 45 percent of the girls indicated that a participant would learn new things or that the program would be helpful in life or in school.

How did girls respond to "What influence do you think METRO has had in your life?" Twenty percent of the girls responded that they "learned to have faith in myself, learned what I really am," and 12 percent responded that the program was a "great influence, glad to have had a chance to come." Six of the fifteen girls in one eighth-grade class viewed the influence as "math class, helped in studies."

Although seventh graders responded "learned things, made me learn, good learning experience," it was the eighth graders who responded "insight on school, influences courses to take." Perhaps this difference is to be expected. Seventh graders frequently have little choice in the selection of courses, whereas eighth graders often begin making choices that lead to academic or honors tracks in high school. When given an opportunity to make other comments about METRO, slightly more than 20 percent of the eighth graders remarked, "Helps you get prepared for school, learned a lot." However, the most common responses by far were "great program, loved it, inspirational place" and "fun." Sixty-two percent of the girls would like to attend METRO again, whereas 25 percent were unsure. In commenting about their mathematics class, the majority responded positively. To the question "How do you feel about your mathematics class?" 55 percent of the girls reported "liked, fun, loved, ok." At the seventh-grade level, the second most common response category referred to the teacher; at the eighth-grade level, the second most common category referred to learning or understanding.

Some differences between seventh and eighth graders were evident when they compared METRO's mathematics course to their regular school class. Eighth graders saw METRO as explaining, helping, or more interesting than school. Seventh graders made more negative comments about either METRO or school or made comments about the pacing and homework involved in the program. Comments about pacing and homework are likely a reflection of the curriculum, which moves quickly and expects students to read on their own and work problems. The fact that the seventh-grade girls are just emerging from an elementary school experience may also explain their lower level of enthusiasm for doing academic work in the summer.

Girls were also asked, "Has the METRO math class changed your attitude toward math? If so, how?" Forty-five percent of the responses were positive, with girls reporting that they found they could learn or that mathematics could be fun. Ten percent of the responses indicated no change in attitude, because mathematics was already liked. The positive comments are important in light of research that indicates that self-confidence and general liking of mathematics are important for continued study of the subject (Sherman 1982; Fennema 1981; Meece et al. 1982).

Eighth grades responded much more positively than seventh graders to the question "Has the METRO mathematics class changed your attitude toward taking algebra? If so, how?" This is likely a reflection of the mathematics curriculum at METRO. The eighth-grade girls study algebra and possibly find that algebra is accessible. Also, algebra may be an option for these participants in the upcoming school year. The seventh graders had less exposure to algebra at METRO, since their curriculum focused more on applied arithmetic with appropriate use of scientific calculators.

Results from Parents

At the graduation ceremony on the last day of the METRO program, parents were asked to complete a questionnaire, one by each family. Thirty-eight responses were obtained. In response to "What qualities, if any, has METRO brought out in your daughter?" 21 percent of the parents responded, "More outgoing," "Speaks up more," "More confidence in herself," "Willing to use her intelligence."

All parents responding to the questionnaire indicated that they would like to see their daughter attend the program again and that they would recommend the program to other parents. Seven of the thirty-eight parents would recommend METRO for "educational reasons, [it] helps girls prepare academically and spiritually."

Results from METRO Graduates

Seven former graduates of METRO, now in high school, were hired to serve as counselors during the 1988 summer program. In early June, before the program began, these girls were interviewed to determine their perceptions of METRO's effects. Clearly these girls constitute a very select sample since they are former participants who are now working for the program. Futhermore, in addition to attending the summer program, all had been involved in the high school and leadership aspects of METRO conducted during the school year. Although their perceptions of the summer program have undoubtedly been affected by participation in these other aspects of METRO, these girls can shed some light on the possible long-term influences of the program.

Girls were asked, "What made you want to attend METRO in the first place?" The top responses, "Meet new people," "Helped me be more friendly," "Made me feel at home," are similar to those expressed by current participants.

The question "What influence do you think METRO has had in your life?" also generated a response similar to that given by current participants. Six of the seven girls responded, "Teaches you to be what you want to be," "You can achieve," "You can go beyond what you thought."

Of interest is the fact that METRO had some impact on the school course choices of these girls. Four of them reported that because of METRO they took more mathematics and science than they had originally planned. METRO was credited with pushing them to achieve.

These girls saw METRO as responsible for building their self-confidence and helping them learn to be whatever they desired. The push to achieve led, in practice, to an increase in the amount of mathematics and science taken by these girls. This is an important influence, considering the role these two areas have in keeping career options open.

Summary and Implications

The METRO Achievement Program is a multifaceted summer academic program for average girls from the inner city who will be entering the seventh and eighth grades in the fall. In addition to helping the girls develop academically, the program strives to help them develop strong character values and self-confidence in their abilities.

Overall, both participants and parents believed that the METRO program was a positive influence. All the parents and almost all the girls would recommend the program to someone else. Although additional long-term evaluation is needed to ascertain long-lasting benefits of the program, the anecdotal evidence, as well as the qualitative evidence collected to date, indicates that METRO can have a positive influence on the academic experiences of inner-city girls.

The results obtained by the METRO program have some broad implications for the educational system. Too often, average students, particularly girls, are allowed to coast through the school system without taking academically challenging courses. The decisions these students make have far-reaching implications on future career and educational opportunities. In inner-city environments, the problem is even more acute because low academic expectations are coupled with high dropout rates. The fact that 85 percent of the new entrants to the U.S. work force of the twenty-first century will be female, minority, and immigrant (Johnston and Packer 1987) implies that we can no longer afford to furnish such students with a curriculum of low expectations. We must begin to provide the needed support in order to reach the stated goals outlined at President Bush's 1990 education summit in Charlottesville, Virginia:

> What our best students can achieve now, our average students must be able to achieve by the turn of the century. We must work to ensure that a significant number of students from all races, ethnic groups, and income levels are among our top performers. (U.S. Department of Education 1990, p. 3)

The METRO program suggests one model for reaching average students, particularly those in an inner-city environment, and upgrading their skills. Intervention in the crucial junior high years can help girls develop confidence in their ability to do mathematics. High expectations by educators, together with some personal attention and encouragement, can lead students to accept academic challenges rather than to slide by with little effort. The confidence that is built in mathematics through the use of applications and technology provides an important building block for continued success.

But such intervention needs to consider more than just academics. As in the METRO program, intervention must consider the whole child. Academic subjects do not exist in a vacuum, and intervention programs for inner-city children cannot afford to ignore the social and economic environments in which these children live. Programs should involve parents as well as help students learn to cope with harmful drug and peer pressures. In this way, children can unify their lives so that academics have a chance to flourish. By helping children at a crucial juncture in their educational experience, we as a society can benefit in innumerable ways.

References

Arias, M. Beatriz. "The Context of Education for Hispanic Students: An Overview." *American Journal of Education* 95 (November 1986): 26–57.

Baron, Reuben M., David Y. H. Tom, and Harris M. Cooper. "Social Class, Race and Teacher Expectations." In *Teacher Expectancies,* edited by Jerome B. Dusek, Vernon C. Hall, and William J. Meyer, pp. 185–226. Hillsdale, N.J.: Lawrence Erlbaum Associates, 1985.

Brush, Lorelei R. *Why Women Avoid the Study of Mathematics.* A longitudinal study for the National Institute of Education. Final Report. November 1979. (ERIC Document Reproduction Service ED 188 887).

Campbell, Patricia B. "What's a Nice Girl Like You Doing in a Math Class?" *Phi Delta Kappan* 67 (March 1986): 516–20.

Chicago Tribune Staff. *Chicago Schools: "Worst in America"—an Examination of the Public Schools That Fail Chicago.* Chicago: *Chicago Tribune,* 1988.

Eccles, Jacquelynne, and Allan Wigfield. "Teacher Expectations and Student Motivation." In *Teacher Expectancies,* edited by Jerome B. Dusek, Vernon C. Hall, and William J. Meyer, pp. 185–226. Hillsdale, N.J.: Lawrence Erlbaum Associates, 1985.

Fennema, Elizabeth. "The Sex Factor." In *Mathematics Education Research: Implications for the 80's,* edited by Elizabeth Fennema, pp. 92–105. Alexandria, Va.: Association for Supervision and Curriculum Development, 1981.

Johnston, William B., and Arnold E. Packer, eds. *Workforce 2000: Work and Workers for the Twenty-first Century.* Indianapolis: Hudson Institute, 1987.

Isaacs, David. *Character Building—a Guide for Parents and Teachers.* Dublin, Ireland: Four Courts Press, 1984.

McBay, Shirley M. "Education That Works for Minorities: An Action Plan for the Education of Minorities." In *Making Mathematics Work for Minorities: A Compendium of Program Proceedings, Professional Papers, and Action Plans,* pp. 30–33. Washington, D.C.: National Research Council, Mathematical Sciences Education Board, 1990.

MacCorquodale, Patricia. "Mexican-American Women and Mathematics: Participation, Aspirations, and Achievement." In *Linguistic and Cultural Influences on Learning Mathematics,* edited by Rodney R. Cocking and Jose P. Mestre, pp. 137–60. Hillsdale, N.J.: Lawrence Erlbaum Associates, 1988.

Maritain, Jacques. *Education at the Crossroads.* New Haven, Conn.: Yale University Press, 1943.

Matthews, Westina. "Influences on the Learning and Participation of Minorities in Mathematics." *Journal for Research in Mathematics Education* 15 (March 1984): 84–95.

Meece, Judith L., Jacquelynne Eccles Parsons, Caroline M. Kaczala, Susan B. Goff, and Robert Futterman. "Sex Differences in Math Achievement: Toward a Model of Academic Choice." *Psychological Bulletin* 91 (1982): 324–48.

Merideth, Charles W. [Banquet Address] In *Making Mathematics Work for Minorities: A Compendium of Program Proceedings, Professional Papers, and Action Plans,* pp. 56–65. Washington, D.C.: National Research Council, Mathematical Sciences Education Board, 1990.

METRO Achievement Program. Program literature, summer 1987. Chicago: The Program, 1987.

Meyer, Margaret R. "Gender Differences in Mathematics." In *Results from the Fourth Mathematics Assessment of the National Assessment of Educational Progress,* edited by Mary Montgomery Lindquist, pp. 149–59. Reston, Va.: National Council of Teachers of Mathematics, 1989.

National Council of Teachers of Mathematics. *Curriculum and Evaluation Standards for School Mathematics.* Reston, Va.: The Council, 1989.

Nelson, Barbara Scott. "Welcome: Perspective on the Issues." In *Proceedings from the Operation SMART Research Conference,* pp. 3–6. Indianapolis: Girls Clubs of America, National Resource Center, 1987.

Nicholson, Heather Johnston. "But Not Too Much Like School: Case Studies of Relationships between Girls Clubs and Schools in Providing Math and Science Education to Girls Aged Nine through Fourteen." Paper presented at the Operation S.M.A.R.T. Research Conference, Girls Clubs of America, Inc., October 23–25, 1986.

Ogbu, John U. "A Cultural Ecology of Competence among Inner-City Blacks." In *Beginnings: The Social and Affective Development of Black Children,* edited by Margaret B. Spencer, Geraldine K. Brookins, and Walter R. Allen, pp. 45–66. Hillsdale, N.J.: Lawrence Erlbaum Associates, 1985.

Scott-Jones, Diane, and Maxine L. Clark. "The School Experiences of Black Girls: The Interaction of Gender, Race, and Socioeconomic Status." *Phi Delta Kappan* 67 (March 1986): 520–26.

Sherman, Julia A. "Mathematics the Critical Filter: A Look at Some Residues." *Psychology of Women Quarterly* 6 (Summer 1982): 428–44.

Silverstein, Barry, and Ronald Krate. *Children of the Dark Ghetto: A Developmental Psychology.* New York: Praeger Publishers, 1975.

Slaughter, Diana T., and Edgar G. Epps. "The Home Environment and Academic Achievement of Black American Children and Youth: An Overview." *Journal of Negro Education* 56 (1987): 3–20.

Stiff, Lee V. "African-American Students and the Promise of the *Curriculum and Evaluation Standards.*" In *Teaching and Learning Mathematics in the 1990s,* edited by Thomas J. Cooney, pp. 152–58. Reston, Va.: National Council of Teachers of Mathematics, 1990.

Syron, Lisa. *Discarded Minds: How Gender, Race and Class Biases Prevent Young Women from Obtaining an Adequate Math and Science Education in New York City Public Schools.* Prepared for the Full Access and Rights to Education Coalition and the Center for Public Advocacy Research. New York: The Center, 1987.

Thompson, Denisse R. *METRO Achievement Program: Summer 1988 External Evaluation Report.* November 1989. (Eric Document Reproduction Service ED 317 651).

U.S. Department of Education. *National Goals for Education.* Washington, D.C.: The Department, 1990.

Usiskin, Zalman, James Flanders, Cathy Hynes, Lydia Polonsky, Susan Porter, and Steven Viktora. *Transition Mathematics.* Chicago: University of Chicago School Mathematics Project, 1986. (A revised version of this text is published by Scott, Foresman & Co., 1990.)

Part 4
Changing How Students Learn

The transition to the learning experiences envisioned the NCTM's *Standards* begins in the so-called traditional classrooms. Teachers are attempting with success to use a variety of learning strategies, some of which are addressed in the chapters in this section. These approaches range from adjusting traditional classroom methods, as related in MacDowell's paper, to very innovative approaches, such as the one described by Healy. The other chapters, by Bebout and by Davidson and Hammerman, discuss two factors related to reaching all students—using student-constructed problems and working with heterogeneous groups.

14

Homogenized Is Only Better for Milk

Ellen Davidson
Jim Hammerman

Differences are everywhere. Different races, ethnicities, and genders. Different abilities and skills. Different preferences and likes. Whether it concerns who we are, what we do, who or what we like, or how we think, differences abound. There's no avoiding them.

Our job as educators is to discover how to take advantage of these differences. At The Phoenix School, we use cooperative, heterogeneous, small-group work and a constructivist framework for education to get the most out of these differences. Each piece—cooperation; heterogeneous, small groups; and constructivism—contributes an important element to the success of our work. Together, they provide a synergistic set of tools that help us create powerful educational experiences.

Who We Are

The Phoenix School is different from most educational settings in a number of ways. We are essentially an urban one-room school. We have one mixed-age classroom with children ranging from six to eleven years of age. Students come to us from urban and suburban homes and from primarily working class and middle class backgrounds, although some of our students have been very poor. The student body is a mix of white, African American, and Hispanic children from Christian, Muslim, and Jewish religious backgrounds.

Students come to us for very different reasons. Some are especially advanced and need to be challenged. Some come because they've had behavioral or learning problems in other schools. Some come as a transition

This chapter is adapted from an article by the same title published in *Democracy and Education* ([Ohio University] 4 [Fall 1989]: 17–22).

from home schooling. Some are seen as being a bit "odd" and have been ostracized in other settings. Some are just average children who come because their parents want small classes, individualized attention, and parental involvement. One things for certain—homogeneity has never been our strong point.

But, as we've said, we think heterogeneity is quite a strong point indeed. We're not going to claim brilliant foresight or insight, although our reflective intuition has served us well. Most of what we've done with children—including mixed-age, heterogeneous, cooperative grouping and access by younger children to the more sophisticated work being done by their older peers—we've done out of a combination of the requirements of our situation and a strong sense of values. We've found that these practices work well for children and are worth implementing in many different kinds of classrooms.

We set up our language arts and mathematics groups to be as academically homogeneous as possible within this setting, although our version of "homogeneous" includes students at several grade levels within any group. For social studies and science we typically have two ways of grouping students, one in which students are mixed across our entire age range and another in which they are grouped more or less by developmental level. Again, "homogeneity" is a relative concept.

Rather than wish for the Holy Grail of "a homogeneous class," we've chosen to embrace heterogeneity and take full advantage of it. Our mathematics class is a good example. The class has students ranging in age from seven to eleven, who would be working at about third- through seventh-grade levels in standard mathematics textbooks. We share the job of teaching this class, each of us alternating days in school with days at home caring for our children. Although we talk a lot at home about the class, we each teach it on our own.

What We Do—*Constructivism*

Our commitment to constructivism is an important part of what makes working with such a wide range of students possible. Constructivism is the belief that students learn by exploring, discovering, and constructing for themselves the concepts that are important. They can learn only what they are ready to learn, though we can facilitate that process by presenting them with problems that will challenge their current assumptions and help them build new concepts. We strike a delicate balance between responding to the interests of, and tangents taken by, students as they explore—their way of telling us what is the next appropriate subject matter—and our own sense of the "big ideas" that we want them to address during the year. But more on this later.

Because students construct conceptual understandings from their own experiences, we need to start there. Most often, we start by posing a problem realm in which our students can explore. This can take the form of real-world problems, games, or even the exploration of patterns in relatively abstract operations. It's better still, however, if these problems arise out of the everyday experiences of our students in the classroom. For example, students can get a real taste of the meaning of division and remainders when sharing any classroom supplies—22 students carving 5 pumpkins for Halloween see that they need 4 or 5 students in each group, not 4 r. 2 or 4 2/5.

Some of our best experiences come when students spontaneously notice something mathematical in otherwise unrelated classroom activities and we decide to pursue it. For example, as we counted students in attendance in our whole-group gathering time in the morning, students found that sometimes they could make equal-sized groups and count by 2s or 3s or 5s but that at other times they could not. On some days, they could group themselves in a wide variety of ways, such as when there were twenty-four students in the class. They wondered what was special about these numbers—both the ones that didn't let them make groups and the ones that allowed for lots of possibilities. This led to an exploration of prime and composite numbers in mathematics class.

Students often use manipulatives and diagrams to help them represent concretely what is going on in the problems they are solving. In this way, they begin to move from concrete situations to more abstract representations of the processes, and they make better sense of what they are doing at the same time. Once the processes make sense to them, but not until then, children work on developing abstract, written symbols to represent their thinking. In part, this is to help them communicate their thinking to a wider audience, but it is also to enable them to use the power of abstractions. (It's hard, after all, to keep a bag of base-ten blocks in your back pocket when you go to the supermarket.) Eventually, they can work problems entirely symbolically—mentally, on paper, or with a calculator—but they understand each step and see connections to the more concrete processes they have used. Because they are developing number sensibilities and estimating skills, they can also check to be sure that their answers make sense—a crucial skill, according to the National Council of Teachers of Mathematics's *Curriculum and Evaluation Standards* (NCTM 1989).

So how does constructivism help? In part, it means that we know that students will be learning material that is at the appropriate level for them—in fact, they can't do otherwise. Because of this, our job is to design "rich" materials and activities—those that will allow for exploration on a variety of levels. An example may help illustrate what we mean.

During a study of factors and multiples that came out of our exploration of primes, we had the children play the factor game (fig. 14.1, modified from Fitzgerald et al. [1986]). We noticed that some students used this game primarily as a multiplication-fact drill, others developed sophisticated strategies for winning the game, and still others constructed number theories about factors. Most students did at least a little bit of each kind of learning. The game was interesting and motivating for all children in the group. They built on the knowledge they already had to explore their own next appropriate ideas and skills, although what was appropriate was different for different children.

Factor Game

Rules: Player A chooses any number and covers it. Player A gets that many points. Player B covers all the uncovered factors of that number and gets all those points. Then they switch, with player B choosing the number and player A getting the factors. If a player chooses a number with NO uncovered factors left, the player gets that many points but loses her next turn. Play continues until there are no more uncovered numbers that have any uncovered factors left on the board.

1	2	3	4	5
6	7	8	9	10
11	12	13	14	15
16	17	18	19	20
21	22	23	24	25
26	27	28	29	30

Fig. 14.1

Because the children end up working on the skills and understandings that are appropriate just for them, this method of teaching is successful even with very mixed groups. Like good literature that can be understood on a number of levels, good teaching should be accessible to children who are learning very different things.

The factor game provided just this kind of rich experience for our students. For example, Adam was a seven-year-old with a very mixed set of mathematics skills. His strategic thinking was amazingly good. Yet his understanding of computational algorithms was quite poor. While they were playing the Factor Game, we asked the children to develop a strategy for the "worst first move." We thought (as did the authors of our sourcebook) that the "right answer" was an abundant number—that is, a number whose factors add up to more than the number. Except for Adam, all the mathematics students who tend to be sophisticated in their analysis of strategy came up with this expected answer. Adam, however, claimed that by choosing 1 as your first move, you'd get only 1 point and you'd lose your next turn, thus giving your opponent an advantage, even though on the first round the opponent gets nothing. Adam explained his theory and we tested it in class. This proved thought provoking to all of us, including the teachers.[1]

However, even though he knew all his number facts perfectly, Adam was one of two students having the most difficulty understanding computational algorithms. In this area he was still working entirely with base-ten blocks.

1. We assume that only player 1's first move is bad, that she is thereafter playing to win, and that player 2 is playing to win throughout. The scores for a thirty-number game board appear in the chart in the footnote on the next page.

Our constructivist approach ensures that we are meeting the needs of all the children in our classroom. Meeting their needs means challenging them just as diligently in their areas of strength as in the areas where we might typically say they need more work. Adam, for example, is being challenged strategically as well as computationally. This is also true for the other children, even though they have different combinations of strengths and weaknesses. Yet it would be impossible to do these kinds of activities with a group of children of any size if we didn't add other elements. Cooperative small-group work is one of the primary additional elements.

What We Do—*Cooperative Small-Group Work*

Students in our class work in cooperative groups to solve problems. Together they devise methods to approach problems, with each child explaining her thinking along the way and building on the thinking of others. We also build in interdependence, making each student in a group responsible for the others. For example, students work together on a single worksheet and then sign it when they are done to show that they both agree with and understand all the work on it. (We often photocopy these after they are completed so that each student has one to take home.) Once the sheet is signed, all students are responsible for everything on it, and we may ask any of them to explain the thinking of the group or each of them to represent the group in a new group made of students who got different answers (Schniedewind and Davidson 1987; Johnson and Johnson 1984) Although we cannot work directly with every group at once, having the students work in small groups with peers means that they can give each other feedback, ask each other questions, and try to communicate their understandings and strategies. Students clarify their thinking and deal with misconceptions, and they are never stuck just sitting and waiting for a teacher.

Children often learn the most from each other. We want children not only to understand a concept but to be able to explain their thinking. Because

	Our Answer		*Adam's Answer*	
	Player 1	*Player 2*	*Player 1*	*Player 2*
Round A, player 1	24	36 (1, 2, 3, 4, 6, 8, 12)	1	0
Round A, player 2	5	25	5	25
Round B, player 1	27	9	skipped	skipped
Round B, player 2	13	26	12 (3, 9)	27
Total	69	96	18	52
	Player 2 is ahead by 27.		Player 2 is ahead by 34.	

students share a worksheet when they work in twos or threes, they must find ways to come to agreement about the answers. When they disagree about an answer or a method for solving problems, they are motivated to talk about those differences, explain their thinking along the way, listen to each other, and try to convince each other. This focus on the reasons behind answers gets students thinking about what mathematics means and helps them see themselves as mathematical meaning makers. By talking with, and listening to, each other, students can solidify or change their thinking, together coming to a new synthesis that will be more robust than the thinking of any one of them alone.

But it's not just that our children work in small cooperative groups. An important element of our classroom is that these groups are often markedly heterogeneous.

Thoughts on Homogeneous Groups

We *like* working with this wide range of children all together. They stimulate each other, understand each other, and help each other. Such a wide range gives students easy access to varied placement in different content areas and to movement among groups. Yet when we read about those who prefer homogeneity, we like to try to understand what advantages they see.

It seems to us that teachers who want tracking, or some other version of homogeneous grouping, want it because they think that it works better for students and because they expect that their jobs will be easier. The theory is that by getting a group of children who are "at the same level," a teacher will find that his or her teaching can be simplified because it need be directed only at that level. Teachers will have less preparation work to do, and discussions will always be "right on target." Students will neither be bored, if the lesson is on something they already understand, nor frustrated, if it is too difficult. In theory, homogeneous grouping makes everybody happier.

There are a number of problems with this approach, however. First, our experience is that people don't *come* in homogeneous groups, no matter how hard we try to make them do so. Even if students test at the same level in a specific subject, they differ in learning styles, backgrounds and experiences, abilities in other subjects, family situations, moods, and so on. Even if we could find an entire class of children who tested at exactly the same level in mathematics, there would be some who got that score because they were intuitive problem solvers, others because they were fast at computations and therefore got a lot done, and still others because they were slow but very accurate in their work.

And one cannot even predict a definitive sequence in acquiring particular skills or knowledge. As we saw with Adam and the factor game, his knowledge of number facts and of the multiplication algorithm was quite weak even though his understanding of number patterns and strategies was very strong. In a typical classroom, his skills would likely have been lost because of his computational weaknesses.

Slavin (1988) cites research by Goodlad to show that attempts to create two or three homogeneous classes at a particular grade level reduce total variability by only 7 percent and 17 percent, respectively. Such a reduction is not likely to make teaching much easier. It does create, however, the perception that there *are* "low" and "high" classes, with the inevitable stigma and effects on student self-esteem that go along with that perception (Oakes 1985).

A second problem is that when teachers *think* that they are teaching a homogeneous group, they don't take advantage of the diversity of skills and knowledge that students bring to the subject being studied. They also don't sufficiently account for differences in students' learning processes. Students get a chance neither to work from their areas of strength nor to learn from the diverse approaches that others bring to a discussion. It seems that it *ought* to be reasonable to expect all members of a class that is designated "homogeneous" to learn the same content in the same way and at the same speed. Unfortunately, these kinds of assumptions can be quite dangerous.

What We Do—*Heterogeneity*

Not only can heterogeneous work be made easy, but also it can enrich children's learning experiences in essential ways.

As students discuss their different viewpoints with one another, they can become either more enlightened or more confused by these explanations, both of which we think are fine, for different reasons. Sometimes another childs explanation helps to clarify something better than our explanation as teachers ever could. At other times, the explanation serves the important purpose of throwing a wrench into a fragile understanding, forcing a student to rework it into a more solid form. As teachers, we try to facilitate these discussions, often asking students in different groups and with different opinions to talk with each other until they can agree. Encouraging students to explore the *different* opinions and understandings that arise is an important tool for increasing learning.

Vygotsky (1978) alludes to the importance of this kind of work through his concept of the Zone of Proximal Development (ZPD). He distinguishes between those skills that a child can demonstrate solidly on her own, those that she can demonstrate only with help, and those that she cannot demonstrate at all. It is the middle range (the ZPD) that provides the most potential for growth. And it is often those who have just recently conquered a concept—namely, other children—who can best speak to the changes in understanding that have occurred.

We've discovered that this middle zone often extends both further up and further down than we might expect. On the upper end, children can typically grapple with more complex ideas than they are usually given credit for. They can become familiar—even friendly—with concepts that they might not even address until much later in traditional classrooms. On the lower end, they too often haven't solidly understood concepts that have been explored, even very thoroughly, and need to remind themselves in new situations of what both we and they know they know. Our goal, then, is not to have children reach "closure" on a particular concept; rather, we want important ideas to remain as open questions, so that students' understanding can evolve into more robust forms. Our job is to present an alternative perspective—often a less developmentally advanced one—to try to "trick" the students into questioning their answers and to get them to rethink their conclusions. By encouraging children to explain their thinking, we help them solidify their understanding. As they express their ideas and answer other children's questions, they make deeper connections, develop new insights, and further convince themselves of the truth of their understanding.

When we work with children in heterogeneous groups, our discussions tend to focus at somewhere above the middle of the group's abilities. This means that work is challenging for the middle and upper range of students. Because our content is "rich," there is appropriate work for "lower" students to do as well. But it also means that the "lower" students are exposed to more sophisticated ideas than they would be in a homogeneous classroom. Often it turns out that children whom we might perceive as "lower" can understand content that is more complex than we would expect them to be able to handle. They become intrigued by the ideas, even friendly with them, and so are motivated to explore even difficult concepts in a way that is appropriate for them.

In schools with tracking, students in "lower" ability groups are not exposed to content that teachers perceive as too hard for them. Therefore, as students have developmental spurts, they may lose the opportunity to move ahead quickly because they're not exposed to sufficiently challenging content. They also don't hear more sophisticated discussions about a topic, and thus lose the positive modeling that that would bring.

This kind of learning, through unplanned exposure to more sophisticated ideas, happens frequently with the students in our classes. For example, two eight-year-old girls—Joanne and Rebecca—were discussing a story they were reading that took place at the turn of the century. They wondered why the Russian peasants in the story were washing their clothes in the river. They concluded that washing machines hadn't been invented yet and that the family didn't want to waste the running water from the sink (which they assumed the peasants had). Jorge (age six), who was listening in, protested that washing machines must have been invented. He had heard that the family was coming to America on a ship, and if ships were invented, "then machines were invented." This got all the children into a good discussion about inventions and changes through history. Not only was Jorge challenged in his thinking about history, but the two girls needed to deal both with communicating their *understanding to him and with reassessing some of their own misconceptions.*

We worry a bit that by working slightly above the average our "lower" students may not get the benefits that come with figuring something out for themselves. We wonder if they'll essentially tag along on other children's learnings and never feel the power of making their own discoveries. Yet the children remind us that this isn't so.

The children were exploring the concepts of area and perimeter by arranging small square tiles in different rectangular shapes. Over the course of the first week of the unit, a number of children discovered that the area (the number of tiles) always turned out to be the number across multiplied by the number long. This was an exciting discovery that, after some discussion, all the students used regularly. Some children used it because they clearly understood how the concept of multiplication applied to the tiles. Other children used it because it had turned out to be true often enough for them to be convinced it was likely always to be true.

Erin was one of the children who consistently used this technique in school and seemed comfortable with it. However, we noticed one day when she was doing her homework that she was counting up the number of tiles after having multiplied. We asked her why she was doing this. She answered, "I know you can multiply them. But I just want to be sure it will come out right."

It became clear to us that even though she "understood" the new concept in the classroom, she didn't fully believe it. This reminded us that having heard another student's explanation was not going to deprive Erin of the chance to learn a concept for herself. In fact, she wouldn't learn it until she was ready. And when she is ready, she'll still have the opportunity to feel the joy of the "Ah ha!" experience of discovery.

Appropriate Curriculum—Different Learning from the Same Problem

Do we really expect children at lots of different levels to do the same kind of thinking? Of course not. But there are times when it *is* appropriate to give them the same work.

Tahira and Sakeera—who who were generally the most advanced students in computation—were doing long division with two-digit divisors and five-digit dividends. In an extracurricular summer program they had learned long division by memorizing the algorithm. They were reasonably accurate at applying the algorithm, but they occasionally made a number of different types of computational errors. Because of these errors, their remainders would sometimes come out to be larger than their divisors. However, they didn't see this as a problem.

In deciding what to do with our heterogeneous group, we could have created a minilesson with just the two of them, or we could have designed independent work that would have helped them investigate this issue more fully. Instead, we decided that exploring the idea of the meaning of a remainder in division would be appropriate for all the children, even those who were just doing short division with small numbers and still using blocks.

We began the discussion by asking the children what the largest remainder could be in the problem 23 759 ÷ 46. Suggestions included the following: "Probably not much more than 100," "47," and "There's no way to tell."

We then gave them the problem 643 ÷ 7 and asked the same question. This time, though some of the answers were similar to those from before, Alan said, "6."

Then Carl interjected, "Oh, oh! In the first one, the largest remainder could only be 45." Most of the children looked mystified by this sudden insight.

James concentrated for a bit and then said he thought he had a rule about remainders that might be true. "The biggest the remainder can be is the number 1 less than the divisor," he said earnestly. He wasn't very sure of this, however.

We asked whether people thought that James's rule made sense to them. Joanne blurted enthusiastically, "Oh, if the remainder was bigger than the divisor, that would mean you could have more things in each group!"

"Or you could have more groups," said Anne.

This led naturally to a whole-group exploration of the two types of division and what remainders could mean in each.

In this incident, some of the beginning dividers were dealing with basic concepts of division before they had much of a sense of the computational process of division. Others were grappling further with some of these basic concepts while they were also developing their skills at the process. Still others were going back to reinforce their understanding of the concept, though their facility at the algorithm was good. Yet despite their differences in skill level, the same activity was appropriate for all these students.

It's important to note that even when students are addressing the same problem, we, as teachers, can help each of them focus on aspects of the problem that are individually appropriate. When we approach a group of students, the types of questions that we choose to ask will vary depending on the skills and the sophistication of understanding of the individual students. We expect each student really to understand the worksheet, but the exact nature of that understanding will differ from student to student. This enables us to use the same set of problems to teach students at a variety of levels.

Appropriate Curriculum—Listening to Students

But there is another tension that we struggle with all the time in creating a constructivist curriculum for a heterogeneous group of students. This is the tension between the course of study being generated from the day-to-day interests and concerns of students and that being generated from our own broader vision of what we'd like students to address. In order to better understand how we work with this tension, we will describe some examples of how our curriculum develops in this context.

The balancing that we do with these two forces expresses itself at a number of levels. On a broad level, we sometimes shuffle the major topics of our curriculum, or even do things we hadn't planned on, because of student interest. For example, as we have described above, it was the students' interest in primes—expressed as noticing a pattern with certain numbers while counting and grouping students during gathering time—that got us into that topic, although we hadn't originally planned on addressing it at all. This led naturally into a study of factors and multiples, which proved extremely useful later in the year as students explored equivalent fractions and the finding common denominators with the addition and subtraction of fractions. Though we hadn't planned on this sequence, we would probably choose it intentionally in future years.

On a smaller scale, this tension finds a different balance within the context of a specific unit. That is, the original problems that students address come from us, but later work arises from the questions, concerns, and less-than-solid conceptions of our students.

In our fractions unit, we begin by asking students to work with manipulatives to name fractions and then to discover rules about how to compare fractions. We then want them to focus on the addition and subtraction of fractions, with the specific intention that they should invent rules about common denominators. To this end, we provide both word problems describing realistic situations in which fractions would be added or subtracted and some purely numerical computation problems so that students can focus more explicitly on the numerical patterns and rules. Through all this, we try to be sure that the numbers involved can be represented by the available manipulatives.

Soon, one of two things happens: (1) students become facile at performing concrete manipulations to solve problems but do not reflect on or generalize their actions to develop rules; or (2) students begin to generate their own rules for describing the manipulations they are doing with concrete materials. In either instance, we challenge students by asking them to do problems that cannot be represented by the available manipulatives. The first situation pushes students to begin to notice the patterns of what they are doing entirely with manipulatives so their concrete actions can begin to become more abstract rules. The second situation pushes students to check the validity of their newly forming rules in the context of more abstract problems.

In either situation, our curriculum follows our students' thinking. That is, in the work we present to them each day, we respond to what our students are thinking. Sometimes this responsiveness is explicit—worksheets restate for the students the tentative rules they have been formulating on previous days and present new problems that will help them examine the strengths and weaknesses of their rules.

An example of this is the worksheet entitled We Are the Rulers of the Rules (fig. 14.2). Students worked with fractions that are not directly representable with the manipulatives available but that can be represented by imagining modifications to the manipulatives or by using diagrams.

By paying attention to our students' ideas and difficulties, we discovered—or rather relearned—that common-denominator problems come in a variety of types:

1. The denominators are already the same.
2. The common denominator is one of the given denominators.
3. The common denominator is the least common multiple of the denominators but not their product.
4. The least common multiple of the denominators is just their product.

Though these distinctions make sense, their importance to students was not clearly apparent to us before we explored the topic in our classroom. Rather, we learned as we worked with students that they needed to deal with all these types of problems in order to construct rules that would work generally. Thus we tried to include all, or at least a variety, of these types in our problems for students. Listening to our students improved our teaching.

We Are the Rulers of the Rules

We've been exploring different ways to add and subtract fractions. Sakeera suggested a rule that said:

"If the denominators are the same, keep that denominator and just add the numerators."

We then added an addendum to this rule that said:

"If the denominators aren't the same, make them the same."

We were wondering if we could always do this. Anne made the following suggestion:

"You can make the denominators the same by finding the Least Common Multiple (LCM) of the denominators."

She had two ways to do this. One was to list multiples of each denominator until you found one that was on all the lists. The other was to use factor trees to help you find the LCM.

Try the following problems to test these rules. When you are done, decide whether or not any (or all) of these rules work.

$3\frac{4}{7} - 1\frac{6}{7} =$ $\frac{5}{6} + \frac{4}{9} =$

$8\frac{2}{3} + 6\frac{7}{15} =$ $8\frac{3}{4} - 4\frac{1}{6} =$

In the space below, write whether you agree with Sakeera's and Anne's rules. Do you think they will always work, sometimes work, or never work? Why? You MUST explain your thinking.

A number of people think that sometimes you can add fractions by adding the numerators and adding the denominators. Some people remember having done this successfully, but no one was able to come up with examples in class when we talked about it.

Some times that people thought this theory may apply are—

- when the numerators are 1;
- when the denominators are the same as each other.

Check out these examples and any other ideas you may have about when this method applies. When you are done, write about whether or not you think this method will ever work. If so, explain what conditions have to be true to make it work. If not, explain what's wrong with the method that makes it never work.

Fig. 14.2

By listening to our students we also help them become more confident mathematically. The mathematics that is being made in the classroom is theirs, not ours or a textbook's. It is our *students'* thinking coming out of group discussions that provides material for the next day's lessons. The problems we pick help students focus on the consequences of, and the possible inconsistencies in, their own thinking. Since we reflect both correct and incorrect rules to our students, the process of validating or disproving the rules is one that clearly belongs to the students as a group. As teachers, we don't have to worry about "missing" important mathematics, because the group's heterogeneity assures that *someone* is likely to have invented an important piece of an appropriate system or a piece of mathematics for the whole group to focus on in the next lesson. Giving students back their thinking respects them as thinkers and helps them see that *they* are making mathematics, not learning from authority.

Measuring Success

So how do we know we are successfully putting all these pieces together? First, since we are constantly listening to students to develop appropriate curriculum, we are also getting a very clear sense of what their thinking is and how it changes with time. We get much more information about the subtleties and details of how students think about important mathematical concepts when we ask them to explain their thinking, and especially to argue against ideas they just recently had or could have had, than we possibly could by merely assessing their success or failure at solving a number of problems.

But this is not the only way we measure success. In part, we also measure the success of our mathematics program by our students' attitudes toward mathematics and toward themselves as mathematicians. Our students like mathematics. They see themselves as powerful *makers* of mathematics. And they think mathematics makes sense. These attitudes persist even when students move on to other schools. We hear from our past students that they become very frustrated when teachers tell them to do something "because that's the rule"—they expect to explore mathematics until they understand it.

Because our methods of instruction are so different from those of most schools, we feel it is important for students to take standardized tests to allay fears that "different" means "less than" and to place students academically when they move on to other schools. Our students consistently do well on standardized tests. But also, our students are excited to take the tests at the end of the year and are often able to do problems that we have never addressed in class.

For example, during a discussion of interesting problems from a standardized test, which had already been completed and scored, Alan raised a question about $\frac{1}{2} \div \frac{1}{2}$. He explained that he thought he knew the answer on the basis of his understanding of division, although we had not dealt with either the multiplication or the division of fractions in class. We asked him to explain his reasoning, which led into an exploration of the meaning of the division of fractions. The worksheet in figure 14.3 shows how we structured this exploration to help students move from what they solidly knew about division with whole numbers toward division with fractions. Again, because of our carefully designed structure, different students got different things out of this exploration. For some, it was just a beginning introduction to thinking about what dividing fractions might possibly mean. For others, it was exactly what was needed to help them invent the "flip and multiply" algorithm. For still others, the worksheet helped them make sense of the algorithm, which they had previously learned by rote.

Conclusion

So, is it worth the work? Both of us believe unequivocally that it is, though it is a lot of work. Not only do children learn better as they work intensively together in heterogeneous groups, but their attitudes change as well. In part, they develop acceptance and respect for those who are different from them. Because they need to work well with students of different genders, ethnicities, races, and family backgrounds to succeed, students develop the skills and attitudes needed to do so. Equally important, they develop an appreciation for the differences in thinking that each of them brings to the work. It is primarily through understanding and communicating about these different perspectives that children learn and grow.

A Problem in Parts

A few days ago we were discussing a problem from the standardized tests:

$$\frac{1}{2} \div \frac{1}{2}$$

Alan said he could solve this particular problem without knowing anything about division of fractions. He was using his knowledge of division of whole numbers and decided that the answer had to be 1. This worksheet will give you the opportunity to explore more about multiplication and division of fractions. At the end of the worksheet, we'll come back to Alan's thinking.

For both multiplication and division, we're going to start with whole numbers and then move our way into fractions.

1a. Start with the problem: $15 \div 3$. Write a word problem for this where 3 is the number of groups you have.

1b. Draw a diagram showing how you would solve this problem.

2a. Now write a word problem for $15 \div 3$ where 3 is the size of each group.

2b. Draw a diagram showing how you would solve this problem.

3. Now take the problem $14 \div \frac{1}{2}$. Write a word problem for this and draw a diagram showing how you would solve it. Is $\frac{1}{2}$ the size of each group or the number of groups?
4. Now try $6\frac{3}{4} \div \frac{1}{4}$. Again, write a word problem and draw a diagram showing how you would solve this problem. Is $\frac{1}{4}$ the size of each group or the number of groups?
5. Write a word problem to go with $7\frac{1}{3} \div \frac{2}{3}$. Draw a diagram and solve the problem. Is $\frac{2}{3}$ the size of each group or the number of groups?
6. Write a word problem and draw a picture to solve: 4×5.
7. Write a word problem and draw a diagram to solve: $8 \times \frac{1}{3}$.
8. Do the same thing for $6\frac{3}{4} \times \frac{1}{2}$.
9. Go back to Alan's thinking about $\frac{1}{2} \div \frac{1}{2}$. On the back, explain how you might know the answer to this problem is 1 without having to know how to do division of fractions.

Fig. 14.3

References

Fitzgerald, William, Mary Jean Winter, Glenda Lappan, and Elizabeth Phillips. *Factors and Multiples.* Reading, Mass.: Addison-Wesley Publishing Co., 1986.

Johnson, David W., Roger T. Johnson, Edythe Johnson Holubec, and Patricia Roy. *Circles of Learning: Cooperation in the Classroom.* Alexandria, Va. Association for Supervision and Curriculum Development, 1984.

National Council of Teachers of Mathematics. *Curriculum and Evaluation Standards for School Mathematics.* Reston, Va.: The Council, 1989.

Oakes, Jeannie. *Keeping Track: How Schools Structure Inequality.* New Haven, Conn.: Yale University Press, 1985.

Schniedewind, Nancy, and Ellen Davidson. *Cooperative Learning, Cooperative Lives.* Dubuque, Iowa: W. C. Brown, 1987. (Available from Circle Books, 30 Walnut St., Somerville, MA 02143.)

Slavin, Robert. "Synthesis of Research on Grouping in Elementary and Secondary Schools." *Educational Leadership* 46 (September 1988): 67–77.

Vygotsky, Lev Semyonovich. *Mind in Society: The Development of Higher Psychological Processes.* Edited by Michael Cole, Vera John-Steiner, Sylvia Scribner, and Ellen Souberman. Cambridge, Mass.: Harvard University Press, 1978.

15

They Can Learn

Nora MacDowell

It usually begins at the end of April or the beginning of May. The entire high school mathematics department meets to decide who will be teaching which courses for next year. The department head turns to you and you hear those fateful words, "It's your turn to have the remedial classes." You know the class: the burn outs, the behavior problems, the kids no one else will take, the turned off, the tuned out. You feel your only objective will be self-survival. They won't try any work, won't do any homework, won't even be awake most of the time. After the first wave of panic passes, depression sets in. Summer gives a few weeks of blissful forgetfulness. But then the reality hits—"What am I going to do?"

In our high school, the 5 percent to 8 percent of students who are in remedial mathematics come from various backgrounds with varying skill levels. The one thing they have in common is that they do not demonstrate competency in the basic concepts and skills. The reasons for lack of successful learning are as varied as the students themselves. For some students, a mental block exists with fractions; for others, absence in grade 4 for division by a two-digit divisor left a gap in the learning process; and for other students, a discipline problem made it impossible for them to stay in a classroom long enough to learn any procedure, let alone grasp any concept. Typically, these students are prepared to fail, ready to fight any approach, and more than willing to vegetate or cause enough trouble to be removed from class or suspended from school. The only reasons they have enrolled in any mathematics course are that the school says they have to have so many mathematics credits to graduate and that they aren't of legal age to drop out of school. Their only goal is to reach that magic age or to con some teacher into passing them. Few are actually hoping that this will be the year they finally understand something.

Sound familiar? On the first day of school, these comments are usually heard: "I can't do math." "I hate math." "I'm not doing this and you can't make me." "Dad couldn't do math either." "I'm never going to have to use this." Somehow, these students need to be taught not only the mathematics they will be expected to use but also the self-confidence to keep future doors open to them. And yet by the time they reach high school, they generally have a history of being easily distracted, having a low frustration level, refusing to do homework, and rebelling against absolute teacher control. They are frequently absent (my students average over fourteen days out of school per student per semester) and resist doing makeup work. Is it any wonder, when faced with this scenario, that teachers just want to give up?

These students, however, are reachable—but not necessarily with the traditional classroom approach. They need a structured environment with some flexibility. They need material given in short practice segments with instant feedback to see immediate results. They need specific performance objectives to ensure the confidence of successful and frequent mastery. They need a release in physical terms—movement, a change in task. They need to see that mastery of mathematics and passing the course are not the same. They need to have the possibility of failure taken out of the system just because they can't grasp a concept in someone else's timetable.

I furnish this type of positive environment with an individualized remedial mathematics course based on the positive success of the individual student. It is intended for those secondary school students who have not yet mastered many of the basic concepts and skills. It is comprised of drill and activities that lead to the mastery of these concepts on a step-by-step basis. The course is structured to allow for the greatest growth and flexibility possible. The core of the program is a set of worksheets and activities with diagnostic tests, which act as a foundation on which to build individual instruction and course design.

This individually structured program consists of one hundred units corresponding to specific pupil performance objectives in whole number, decimal, and fraction skills and concepts, along with a few miscellaneous skills. Each unit consists of an explanation sheet, a series of worksheets and activities, a self-test, and two mastery tests. Two diagnostic tests, one without the use of a calculator and one with, in whole numbers, decimals, and fractions, are used to determine if the student needs the computation units. After a prescription is made for each student, he works through the explanation sheet and enough worksheets and activities to understand the objective, checks his own work as he goes along, and then takes the self-test in his first unit. If competency is shown, the student takes a mastery test. A score of 75 percent or better on the mastery test allows that student to move to the next unit. If mastery is not achieved, the student continues working on the material and then retakes the mastery test when greater competency is shown.

My role as teacher in the individualized program allows me the maximum possible student contact. By not having to stay behind a desk or at the chalkboard, I can interact with each student on a daily basis. This interaction may be a pat on the back for a good job, a two-second answer to a question raised, a ten-minute explanation of a process or concept, or even a friendly cajoling. This gives me the luxury of knowing my students, of building their self-confidence, and of being able to give them instant feedback.

Grading in this program is also an individualized process. The mastery test scores are averaged by using the number of units mastered. But this is only one portion of the individualized program. Since this approach also stresses attitudes and classroom decorum, two other grades are also used in compiling a marking period grade. The more objective of these is a daily grade. Everyone is assigned 100 points at the beginning of the marking period. Points are then subtracted for anything that adversely affects a student's daily work. If the school marking period is nine weeks, then 2 points are appropriate each day; on a six-week schedule, 3 points each day. All or any part of these points is subtracted for any violation of classroom rules. Unexcused absences, suspensions, nonwork, or excessive talking can then be taken into account in the grading process.

The third grade is very subjective and involves my assessment of each student's potential and how that student is working up to that potential. Obviously, the ability levels in any group run a wide gamut; as a rule, this applies to the students in our remedial mathematics course. Some students have shown little understanding of most prerequisite concepts, whereas others are weak in only one or two areas. After a period of observing each student, I can intuitively make a judgment on his or her potential. It then becomes a process of evaluating progress against that potential.

For example: If Betty came into class with good computational and calculator skills, she would be expected to cover the purely procedural units with little problem. So if she mastered only the four basic operations with and without a calculator in one marking period, she would not appear to be working up to potential. A grade of 50 percent or 60 percent might be assigned. If Bob, however, came in unable to subtract whole numbers where borrowing is involved yet mastered division of whole numbers with and without a calculator and then proceeded to show he understood equations involving the relationship between multiplication and division, he would be working much closer to potential. A grade of 95 percent or 100 percent could then be assigned.

Each student has to be judged on his or her own merit. After working with these students over the years, I am still amazed at the potential I see. By giving students material at their own level and pace, I am able to see their strengths and abilities breaking through the facade of uncaring. Matching a realistic potential to each student really does become an intuitive process. This grade also puts the student's work into an individual perspective.

These three grades—mastery tests, daily work, and potential—can be averaged to arrive at the final grade for the marking period. All aspects of the student's performance have been taken into account—success, behavior, attendance, and even attitude.

Exams are given at the end of each semester. A student is given one of twenty different exams. The exams are written from material starting with the first unit on whole numbers through twenty different ending units. The student is given the exam that is closest to, but not beyond, his or her progress. This creates an exam as individualized as the program itself.

With the flexible structure of the individualized program, many different learning experiences can be used. Manipulatives, computer work, listening tapes, and small-group work can all be incorporated and aimed at those students who can benefit most. The teacher can vary the activities and projects prescribed for any individual student.

The individualized approach gives much of the responsibility for learning back to the student and gives the teacher the luxury of teaching. Students and parents can no longer use these excuses: "You went too fast." "I thought I knew it." "I wasn't here when you did that." "I just can't do it. SEE!" These rationalizations lose power when students are working at their own pace and at their own level, and the teacher is free to help.

Any program or method of instruction is good only if it fits the students' needs and the instructional goals set by the school and the teacher. These needs and goals are going through a drastic revision, and I have worked to adapt my program of individualization to keep up with those changes. When I wrote my first individualized program for junior high school students in 1973, all emphasis in mathematics was on facts. Drill and more drill was the rule. My second implementation at the senior high school level in 1985 emphasized life skills. The facts and procedures were to be used in those life skills.

But this is 1992. With the adoption in 1989 of the National Council of Teachers of Mathematics's *Curriculum and Evaluation Standards for School Mathematics*, emphasis is changing again toward more reasoning and the use of a calculator to help with computation. My individualized program has been once again revised. I have updated previous units and added new ones to meet the challenges of the next century. A representative sampling of these units appears in figure 15.1.

Sample of Individualized Math Units

(Calculators may be used unless "¢" is shown.)

Whole Numbers and Money Decimals

1. Addition and subtraction facts to three digits—¢
2. Making change
3. Quick math and estimation (×, /)—¢—10, 100, 1000
4. Ratios (using :)
5. Order of operations

Decimals

1. Meaning
2. Quick math and estimation (+, − ,×, /)—¢—moving decimal
3. Scientific notation
4. Equations
5. Solve percent problems by using proportion

Fractions

1. Meaning
2. Order
3. Patterns and sequences
4. Applications
5. Circle graphs

Miscellaneous

1. Simple addition of integers
2. Volume of solids

Fig. 15.1. A representative sampling of individualized math units

Many teachers are getting inconsistent messages regarding the call for change. My home state of Ohio began a ninth-grade proficiency test this year, which must be passed before a high school diploma will be granted. As of this writing, the use of calculators will not be allowed. Anyone wishing a diploma will need to know the basic skills and the paper-and-pencil procedures as well as the problem-solving techniques emphasized on this test. My students will be those who have not yet passed this proficiency test.

When the NCTM Standards were announced, I thought they were the ideal but not really feasible—at least not with some of my students. Then when the

Ohio state legislature mandated a testing program based on the Standards, I panicked. With time, however, I have resolved to, and have even become excited about, making these changes work. Philosophically, I support any and all changes that will benefit the students' moving into our changing society. I am now trying to find the best and most practical way to help those students caught in this transition period. Five or six years from now, when all students have been exposed to the Standards throughout elementary school, the secondary school task should not be as great. The core concepts will at least have been presented to all students. But for now, having come through the traditional and segmented programs, the lower-track student is definitely caught in the middle.

Individualization is my only reasonable way of truly helping these students. I look at all the government and corporate reports on education and wonder where these students fit. We, as teachers, need to do something for them NOW so they will have even a chance to be productive in their futures. If these students can only be instilled with the confidence that mathematics is not beyond them, we have at least left the door open for further education.

Individualization is not a panacea. Some students do not put forth any effort. However, most do respond. When they see that the material is comprehendible and that they can achieve, the majority of students see their first real success with mathematics. It is amazing to watch a face light up as a mastery test is graded and a score of 100 percent is placed at the top. It is exciting to be asked what is wrong with a problem that was missed. It is very gratifying to be asked what mathematics course should be taken next. These are the same students who were convinced that failure was their only alternative in mathematics and accepted their fate. Once students begin to see success, they can care; *they can learn.*

16

Using Children's Word-Problem Compositions for Problem-solving Instruction: A Way to Reach All Children with Mathematics

Harriett C. Bebout

Problem solving is the central goal of the mathematics curriculum, and word problems make up an important part of this goal. *The Curriculum and Evaluation Standards for School Mathematics* (National Council of Teachers of Mathematics 1989) specifically calls for the curriculum to include word problems *(a)* that have a variety of structures, *(b)* that reflect everyday situations, and *(c)* that will develop children's strategies for problem solving (p. 20). Carefully selected word problems and well-developed instructional programs can be used to reach this problem-solving goal.

Using word problems composed by children serves as a means for reaching the central goal of the *Standards* as well as for providing classroom instruction that is diverse and potentially interesting to children. Word problems with a variety of structures can be developed when children are asked to compose problems to match various forms of open number sentences. Word problems can reflect everyday situations when children are asked to compose problems about people and things outside of school. Word problems, too, can be used to enhance the development of specific and powerful informal strategies that represent the structure of various problem types.

This chapter describes an instructional program of word-problem solving that was based on children's original word-problem compositions. The program was carried out recently in two second-grade classrooms of an urban neighborhood school. The chapter begins with a brief description of the school demographics and then proceeds with descriptions of the instructional program. The emphasis in this chapter is on highlighting the features of instruction that were used to reach all children during whole-class instruction.

Students and Their School Setting

Instruction on solving and composing word problems was provided to two classrooms of second graders who attended an urban elementary school in a large metropolitan school district. The school was a neighborhood school that served the families of a low-income urban area. Records from the school indicated that 99 percent of the students were from African American families, 87 percent were designated low income, and 35 percent transferred during the school year; in addition, and similar to performances in many other urban schools, 76 percent of the students scored below the national median on a standardized test of mathematics achievement (Cincinnati Public Schools 1989).

The children in these two classrooms came to instruction with various informal strategies, both appropriate and inapproriate, for solving addition and subtraction word problems. Their strategies were identified during preinstructional interviews in which each child was asked to represent and solve with concrete items various types of addition and subtraction word problems and to compose a word problem for the interviewer to solve. (See table 16.1 for problem types and strategies.) Data from these interviews not only provided a unique set of student-composed word problems for future instructional use but also identified the children who were using inappropriate solution strategies.

Alicia and David were two such children. During their interviews they both used the appropriate elementary strategies of "joining" and "separating from" for the most basic problems, Change 1 and 2, but they used inappropriate strategies on most of the other problem types. For example, on the Change 3 problem about Joe and his pennies, instead of concretely modeling the action described with an "adding on" strategy, Alicia used an inappropriate "extract and add" strategy; in other words, she set out blocks for the numbers stated in the problem, 9 and 15, and then added them together because of the key word *more*. Several other children used this inappropriate strategy to solve the Change 3 problem too, and some used it for the Change 5 and Equalize problems as well. The children who used this inappropriate strategy appeared to ignore the situation described in the problem and to focus instead on matching the numbers stated in the problem to a basic fact.

David too used appropriate strategies for the basic Change 1 and Change 2 problems, but for the other problem types he used an inappropriate strategy that was different from Alicia's. David's predominant inappropriate strategy was to repeat one of the numbers stated in the problem as the answer. For example, on the Change 3 problem David said that 15 pennies was the answer because "that's how many Joe found," on the Change 5 problem he said 13 cars because "that's how many cars were there," and on the Equalize problem he said 8 books because "that's how many Carlos read." David appeared to focus on only a part of each story instead of thinking about the whole situation described.

Table 16.1
Addition and Subtraction Word-Problem Types, Informal Strategies, and Structure-based Number Sentences

Word-Problem Type	Informal Strategy	Corresponding Number Sentences
Change 1	**Joining**	7 + 6 = □
Polly had 7 cookies. Her brother gave her 6 more more cookies. How many cookies did she have then?	Join together sets of 7 and 6 and count; the total.	
Change 2	**Separating From**	17 − 8 = □
Sam had 17 caterpillars in a jar. 8 of them got away. How many caterpillars did he have then?	Make a set of 17, take away 8 of them, and count the remaining items.	
Change 3	**Adding On**	9 + □ = 15
Joe had 9 pennies in his pocket. He found some more. Then he had 15 pennies. How many more pennies did he find?	Make a set of 9, add on to reach 5, then count the items added on.	
Change 4	**Separate To**	14 − □ = 6
Louise had 14 cards. She sent some of them to her friends. Then she had only 6 cards. How many did she send to her friends?	Make a set of 14, take away items to reach 6, then count the items taken away.	
Change 5	**Trial & Error**	□ + 5 = 13
Some cars were in the parking lot. 5 more cars came. Then 13 cars were in the lot. How many cars were in the lot at the start?	Make a set, add on 5, count to reach 13. Adjust the set, add on 5, count to reach 13. Readjust.	
Change 6	**Trial & Error**	□ − 9 = 5
Some kids were in the pool. 9 of them had to go home. Then only 5 kids were in the pool. How many kids were in the pool at the start?	Make a set, take away 9, count to reach 5, and adjust the initial set. Repeat until the final set equals 5.	
Equalize	**Matching**	8 + □ = 15
Carlos read 8 books. Alicia read 15 books. How many more books should Carlos reach to catch up to Alicia?	Make a set of 8 and a set of 15; match the sets in a one-to-one correspondence and count the unmatched items.	

Alicia, David, and some of the other children needed special options during the whole-class instruction. They needed to develop further the appropriate concrete modeling strategies that some of their classmates already had developed informally, and they needed to move along with their classmates in instruction on solving and composing word problems.

Instructional Goal, Content, and Plan

The major goal of instruction was to teach the children to make connections between the informal solution strategies that many children naturally develop and the mathematical forms of open number sentences. In these informal, or invented, strategies most children directly model the action, or structure, of various problem types. These strategies become visible during individual interviews when children use their fingers or other concrete items to represent and solve problems. Children's success in problem solving prior to formal instruction comes from their informally developed strategies.

The initial focus of this instructional plan was designed to capitalize on these insights into problem structure by introducing the representation of problem structure with concrete strategies to some of the children and by reinforcing these strategies in others. The later focus of instruction was designed to help all the children build connections between these structure-based concrete representations and the corresponding structure-based open number sentences (see table 16.1).

Mathematical Content

Several types of word problems were included in this instructional program and were ordered according to level of difficulty. Specifically, as displayed in table 16.1, the instructional problem types included the two elementary addition and subtraction problems (Change 1 & 2), two change unknown problems (Change 3 & 4), the equalize problem (Equalize), and two start unknown problems (Change 5 & 6). During instruction on each of these problem types the children represented and solved first with dramatic reenactments, then with concrete structure-based strategies, and finally with the mathematical symbols in structure-based open number sentences. Conversely, for each of the six different number sentence formats, specifically $A + B = \square$, $A - B = \square$, $A + \square = B$, $A - \square = B$, $\square + A = B$, and $\square - A = B$, the children composed word problems that related the mathematical symbolism to realistic situations that were interesting to them.

The content of these lessons did not displace any of the usual curriculum for the second grade. In fact, the emphasis on relating the concepts of addition and subtraction to situations that were realistic to the children appeared to enhance their understanding of these operations.

Instructional Plan

Each second-grade classroom received fifteen daily forty-five-to-fifty-five-minute lessons. These daily lessons followed the plan for effective mathematics teaching discussed by Good, Grouws, and Ebmeier (1983), including the lesson parts of review, development, and practice, along with a fourth part for the composition of original word problems. Each lesson part presented specific opportunities for including all children in the classroom processes.

Review

A brief review was the opening part of each lesson. The focus was on reviewing the mathematical symbols that were necessary for a complete number sentence. Each child had developed a deck of colored index cards to represent the number sentence elements. As displayed in figure 16.1, these symbol decks consisted of a blue card for the operations with a plus on one side and a minus on the other, a yellow card for the equals sign, a pink card with an empty box to signify the unknown, and white cards for the set of numbers.

During the review the teacher asked the children to recall and volunteer the number sentence elements. Most of the children memorized these parts quickly and readily volunteered, but Alicia, David, and a few of the others needed something to assist them. They were helped by looking at the color-coded cards in their symbol decks. For example, at the beginning of the lesson David and Alicia laid out their symbol decks on their desks, noted the different colored cards, and raised their hands confidently to volunteer the various symbols.

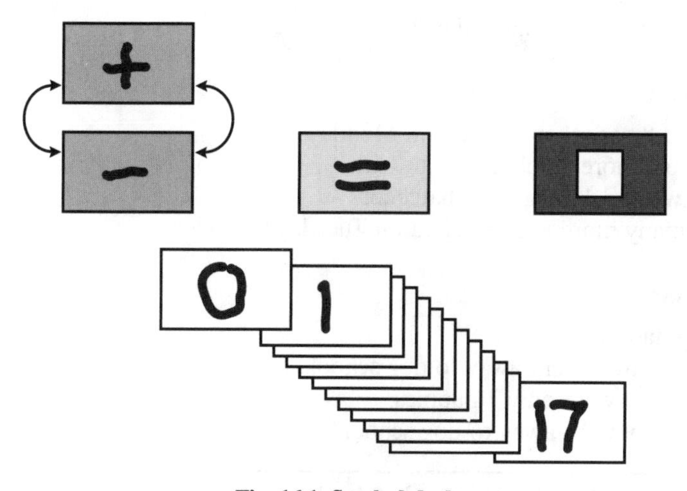

Fig. 16.1. Symbol deck

Development

The development part of the lesson covered a substantial amount of time during the mathematics period. The purpose of this part of the lesson was to emphasize the representation and solution of both new and review types of problems. During this lesson part a total of about twelve word problems were presented on the overhead; these word problems had been selected by the teacher ahead of time from the children's original compositions during the interviews and previous lessons. A typical overhead sheet is shown in figure 16.2. During development the teacher displayed a word problem on the overhead, the student-composer read it aloud, the other children represented and solved the problem at their desks, and then the student-composer represented the problem on the chalkboard and selected a classmate to solve the open number sentence.

David:

We had 14 Transformers.
My brother broke some of them.
Then we only had 4.
How many Transformers did he break?

Donald:

My mom had 17 valentines.
She lost 7 of them.
How many did she have then?

Antonio:

We had 8 eggs that hatched on Monday.
A couple more hatched on Tuesday.
Then we had 11 eggs that hatched.
How many more eggs hatched on Tuesday?

Dwayne:

A dog had 12 puppies.
Some of them went away to dog school.
Then there were only 7 puppies.
How many went away to dog school?

Fig. 16.2. Overhead of development word problems

Certain critical features of the development part of the lesson permitted all children to participate; these features included the preselection of word problems and the promotion of several ways to represent problem structure. By preselecting the problems from previous student compositions, the teacher was able to determine the problem types, the levels of difficulty, and the student-composers who would be the public performers. In selecting problems prior to instruction and preparing them on overhead projectors, the teacher alternated between selecting problems composed by children who needed additional strategy development and problems composed by children who had solid stores of strategies; this alternation between children at different levels kept the whole-class instruction moving at a steady pace.

Another critical feature for involving all the children was the promotion and acceptance of various ways to represent problem structure. The instruction on problem representation covered several lessons for each problem pair. In the initial lessons, representation was carried out by dramatically reenacting the problems, then by using chips or cubes to model the problems concretely, and finally by using number sentences to represent and solve the problems symbolically. Because the representation of word-problem structure was the major focus of the instructional program, representation was emphasized by the teacher as an important step for each problem before the solution; the children became accustomed to explaining or showing how they represented the problem structure before they stated the answer to the problem.

The following classroom episode with Alicia during the development part of the lesson is an example of the teacher's preselection of problems and promotion of different ways to represent problem structure. The teacher had prepared a problem for Alicia that was designed to help overcome her inappropriate "extract and add" strategies and to develop instead strategies that modeled problem structure. Using as the basis a Change 1 addition problem that Alicia had written earlier about her goldfish, the teacher changed the story to a Change 3 format and displayed it on the overhead projector. Alicia read aloud the following story:

> Alicia had 7 goldfish.
> Her family gave her more goldfish for her birthday.
> Then she had 12 goldfish.
> How many did her family give her for her birthday?

Using cubes, Alicia set out 7 cubes to stand for the goldfish, counted on more cubes to reach 12, "8, 9, 10, 11, 12," and then counted these additional cubes to find out how many goldfish she was given for her birthday, "1, 2, 3, 4, 5; 5!" Some of the other children represented the story with the number sentence $7 + \square = 12$ and solved with their counting strategies or number facts. Like Alicia, most of the children were secure in publicly reading, dramatically reenacting, and concretely modeling their own word-problem compositions because they were familiar with the vocabulary and the number domain and were confident that they could reenact or concretely model their stories.

After gaining success through dramatic and concrete modeling of problem types, David, Alicia, and some of the other children gradually moved on to representing the problems with the symbols from their symbol decks. To do this, they laid out the color-coded cards in their symbol decks, selected cards for the operation sign, the equals sign, the unknown, and the given numbers, and then arranged the cards to correspond to the problem structure; when all the elements were in the number sentence, they slipped the card with the solution numeral under the card with the empty box. The final step of writing an open number sentence directly from the word problem was taken by children individually; each of them abandoned their symbol decks when they were certain of the elements necessary for a complete open number sentence.

By having several options for representation, that is, dramatic reenactment, concrete modeling, or open number sentences, all children had ways to succeed in this important step of problem solving. In some lessons Alicia, David, and a few of the other children chose to reenact and concretely model the problems; in other lessons they chose to represent symbolically. But either way they were focusing on problem structure, learning the importance of representation to problem solving, and solving successfully several types of word problems beyond the basic Change 1 and 2 problems.

Practice

The practice part of each lesson focused on the problem types used during the development and usually consisted of an activity sheet with four pictorial word problems. The children were directed to use the pictures to represent the problem, to write a number sentence to match the problem, and then to solve. Children's performances on these activity sheets provided part of the continuous daily progress information to the teacher for planning the subsequent lessons. A typical activity sheet for initial practice on the Equalize problem type is displayed in figure 16.3.

David, Alicia, and the other children worked independently or in student pairs on their activity sheets while the teacher walked among them to observe their progress. After most had finished, the teacher selected children to write the open number sentences on the chalkboard and choose a classmate to solve the sentence; for these public performances the teacher usually selected children who had overcome initial difficulties. For example, the teacher noticed that David had written the number sentence $3 + \square = 9$ and had drawn six more peanuts for Alan in a matching strategy for the peanut story in figure 16.3. David was selected to write his number sentence on the chalkboard and to choose a classmate to put the answer in the empty box.

Name: ______________________

Read the story.
Write a number sentence with a □ to match the story.
Then solve the number sentence.

A

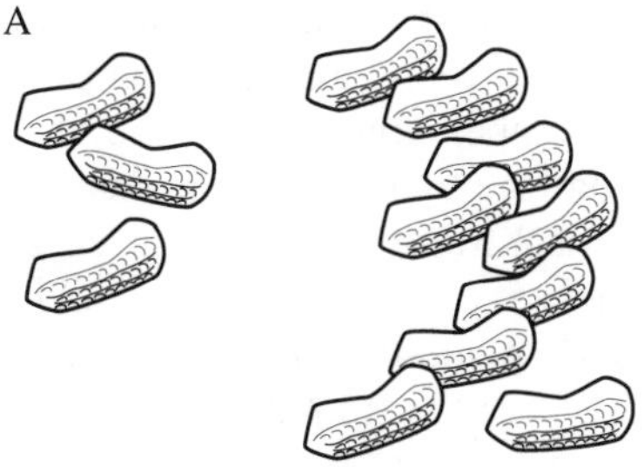

Alan found 3 peanuts.
Susan found 9 peanuts.
How many more should Alan find to catch up with Susan?

____

B

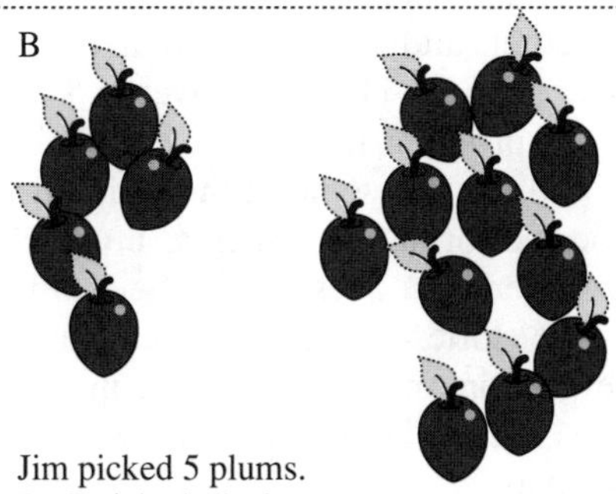

Jim picked 5 plums.
Beth picked 11 plums.
How many more should Jim pick to catch up with Beth?

____

C

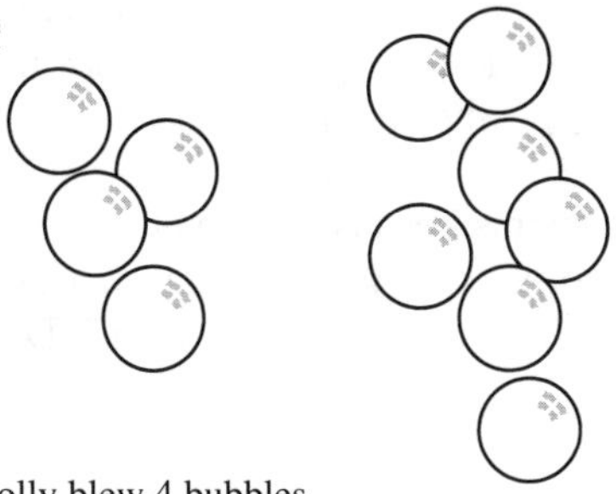

Polly blew 4 bubbles.
Joe blew 7 bubbles.
How many more should Polly blow to catch up with Joe?

____

D

I saw 6 fish in the river.
My friend saw 2 fish.
How many more does my friend need to catch up with me?

____

Fig. 16.3. Sample activity sheet

Word-Problem Composition

In this final lesson part the teacher displayed a pair of open number sentences on the chalkboard and asked the children to compose word problems to match each of the sentences. This lesson part was designed to help children build connections from the opposite direction between the mathematical symbols in open number sentences and their own real-world situations. To do this, the symbolic representations were provided and the children composed word problems to match.

Word-problem composition was probably the children's favorite part. They especially liked joining with a partner and thinking up word problems together, and they knew that their problems might be selected for future lessons. Each child was given a 5″ × 8″ index card and asked to compose a word problem to match each of the open number sentences on the chalkboard. At first some of the children were unaware that word problems had a specific format, or schema, in which numerical information was given and followed by a question; in their initial attempts at composition some children wrote stories about numbers with no concluding question. But after the first few lessons in word-problem solving and with feedback from their partners, all the children learned to write word problems with appropriate word-problem formats.

For example, when Alicia was asked at the end of the initial interview to compose a problem, she related the familiar narrative about Goldilocks and the Three Bears. Later, during instruction she became aware that all the other word problems ended with a question about the numbers in the story. So Alicia changed her story, ended with a question, and matched the story with a number sentence as follows: "There was Goldilocks and there were three bears; so how many were there? $1 + 3 = \square$." She was pleased that she had learned to attach a question to her stories.

David especially liked the Equalize problems because he could use situations that involved his friends and their favorite toys. When the teacher asked the class to write a story to match the number sentence $12 + \square = 17$, David wrote the following:

Dwayne had 12 Transformers.
David had 17.
How many more does Dwayne need to be like David?

A sample of several word-problem compositions to match specific open number sentences are displayed in figure 16.4.

By including this activity in the daily lesson the children learned not only to represent and solve word problems but also to compose word problems that matched specific symbolic representations. This activity provided the teacher with a way to assess children's progress from a different viewpoint and to collect interesting problems for future instruction.

Evidence of Success

Before instruction, in addition to the individual interviews with concrete items, the children were given a paper-and-pencil test in which they were asked to write a number sentence for and then to solve each of the various

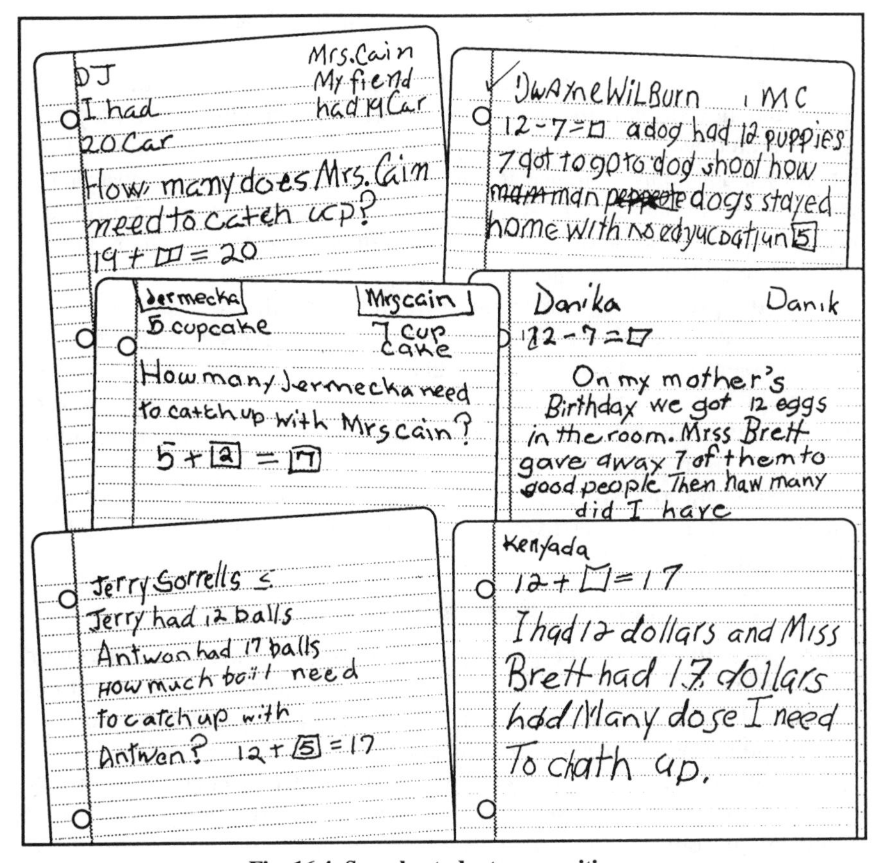

Fig. 16.4. Sample student compositions

word-problem types. Most of the children were very successful on the two most basic problem types, Change 1 and 2, because the structures of these types matched perfectly the traditional number sentence forms of $A + B = \square$ and $A - B = \square$, forms that were familiar to them from their basic-fact study. Some of the children were fortuitously successful on the Change 4 problem and to a lesser extent on the Change 6 problem because these problems too could be represented with the traditional sentence forms.

For example, before instruction, Alicia, David, and many other children were very unsure about how to solve word problems, especially when concrete items were not available to represent the story in the problem. Many of the children resorted to the inappropriate "extract and add" or "extract and subtract" strategies; in other words, they wrote down the numbers given in the problem and selected or tried a plus or a minus sign to determine the answer. By using these inappropriate "extract and add/subtract" strategies, the children were usually successful on the basic Change 1 and 2 problems and fortuitously successful on the Change 4 and Change 6 problems.

For the most part, the preinstructional number sentences that the children wrote did not display the insights into problem structure that they had demonstrated with their informal strategies when using concrete items in the interviews. The only sentence forms that they had seen during instruction were the traditional forms of $A + B = \square$ and $A - B = \square$, and these forms did not correspond to some of their informal strategies, particularly their "adding on" and "matching" strategies.

After the program of instruction on representing the structure of several word-problem types, the children were more successful in solving more of the word-problem types. On the postinstructional paper-and-pencil test, most of the children solved correctly the types of word problems that were very difficult before instruction. The instruction had exposed them to more problem types than the basic Change 1 and 2 types, and they learned that they could represent the structure of different problems not only with their informal strategies but also with number sentence forms that matched the problem structure.

Features Responsible for Success

Certain features of this program in word-problem solving and composition were responsible for the success in reaching all children. These features included the content of instruction, the processes of instruction, and the composition of word problems by the children.

The content of instruction was designed to capitalize on children's naturally developed structure-based strategies. This content led to the development of strategies in some children and reinforced these strategies in others. By organizing the problems according to the order of difficulty and by establishing problem representation with dramatic and concrete modeling, the instructional content capitalized on children's informal understandings.

The processes of instruction were designed to promote success in all children. By identifying children who had not developed all the informal strategies before instruction and by providing them with options during various lesson parts, the teacher kept all the children involved and learning and kept the lesson moving forward at a steady pace.

The composition of word problems to match different forms of open number sentences was an effective way to develop and capitalize on children's informal insights into problem structure. Learning to compose word problems helped the children to connect the number situations of their everyday activities with the numbers and other symbols that they were using in open number sentences. Their compositions added interest, humor, and entertainment to the word-problem instruction and motivated the children for mathematics problem solving. The real-world connections and diversity that these problems brought to the classroom instruction were important features for success in reaching all children. This program of problem-solving instruction that was planned around the children's own word problems was a means for allowing and encouraging all the children in the classroom to participate and be successful in solving addition and subtraction word problem solving.

References

Cincinnati Public Schools. *Rockdale School Annual Progress Report.* Cincinnati: The Schools, 1989.

Good, Thomas, Douglas Grouws, and Howard Ebmeier. *Active Mathematics Teaching.* New York: Longman, 1983.

National Council of Teachers of Mathematics. *Curriculum and Evaluation Standards for School Mathematics. Reston,* Va.: The Council 1989.

17

Equity? We're Just Trying to Survive Here

Christopher C. Healy

When the students passed through my classroom door in the fall of 1987, they expected a year of being stuffed with geometry knowledge. Some looked forward to the meal; others feared it. However, geometry was not prepared for them and served to them that year. Creating and preparing the geometric feast were left to the students.

Every student was in the same foreign environment. The idea of creating their own unique brand of geometry capitalized on their individual differences. They had to learn to work together, to rely on one another, to experiment, to discuss, to present ideas, to compromise, and to really understand the definitions and theories, which the class developed themselves. The students who entered with more mathematical knowledge than others and had had a positive experience memorizing algebraic formulas and using them to solve the equations were shocked. The students who didn't have a pleasant algebra experience, and expected to be unsuccessful in geometry, felt as if they had a chance at success. Even those who had failed a regular geometry class and were repeating the course found hope in this different approach.

They came in expecting lectures, a heavy textbook filled with information, and long reading assignments followed by problem sets. They found none of these. Instead, they were confronted with a teacher who turned over the entire responsibility of learning to the students.

[The following, and later, quotations are representative reconstructions of comments either spoken or written by students.]

> It never seemed to make sense. I do a problem and I know I'm doing it right, but I get a different answer than the rest of the people. I think I just see the world differently than the rest of the people. Give me an easel and a brush, then I can show you what's real and what's not. But taking geometry is going to be a real

disaster for me. In geometry what's right is right, and people who think differently get wrong answers. I may find a dozen wrong ways, but who will care?

—Bernie

I know that geometry is the key to the SAT test. That's why I have to learn everything there is in my geometry class. I've got to be prepared for my SAT's. If I'm not, I won't get accepted to Stanford like I've planned all along. I know I can get an A in the class. I get A's easily, but it's not the A that matters in this class it's the knowledge. I've got to get the knowledge.

—Anna

I guess it had to happen, all summer I've tried not to think about it, but there's no putting it off any longer. School begins tomorrow and that means geometry. I think everyone hates geometry. I didn't do all that well in algebra, so I don't think I have a chance in geometry. All the smart people will understand it right away. Then there'll be me with all the rest, just trying to survive. I think that would be nice . . . to just survive geometry.

—Chris

I know geometry will be easy for me. I've always been good in math. I know I'll have to work and read the book, but I can handle it. Now writing is another thing. I've always had a hard time explaining what I mean in words. It can be all clear in my mind, but when I try to explain it, everything seems to get mixed up. At least math is exact and there are right and wrong answers. In English I only seem to know the wrong answers.

—Edmund

I didn't want to move. I didn't want to come to this school. I don't know anyone, and I don't want to be here. Not only that, but instead of giving me an easy schedule, so I can sort of work into the social scene, they give me geometry. I can handle the other classes, but geometry? I already flunked it at my last school. Give me a break. Instead of being out meeting new people, I'll be stuck at home studying.

—Carmen[1]

How can a geometry teacher hope to teach equitably a class of so many different individuals with such a diversity of attitudes? In my first five years teaching geometry, I learned that the "traditional" method was meeting the needs of fewer than 50 percent of the students, so I began looking for a new approach.

---

1. Information on the students:

Chris—junior girl born in Mexico, lived most of her life in the United States

Edmund—Freshman boy born in Korea, moved to the United States five years before the Build-A-Book class

Bernie—Senior artist born in Puerto Rico, lived most of his life in the United States

Anna—Freshman girl born in the Philippines, moved to the United States early in life

Carmen—Sophmore girl born in the United States

The Background

My search for a geometry class where students experience geometry positively took time. In 1982, I began teaching at the high school level, and in my first geometry class I realized there was something wrong. Not only did I find the failure rate was 50 percent, I also discovered that the students had a very difficult time reading and understanding the textbook. Many geometry textbooks written prior to 1987 seemed to be written with the intent to confuse.

In 1985, as a result of my association with PLUS, I learned that local colleges and industry feel geometry is the most important class a student takes in high school, because "in geometry students learn how to think." (PLUS is an acronym for the Los Angeles project called Professional Links with Urban Schools, one of whose purposes has been to bring together high school mathematics teachers with representatives of colleges and industry, which has allowed each group to see the expectations of the others.)

In the summer of 1987, I attended a four-week institute where I discovered that the available geometry textbooks were similar to the one I used and that the idea of teaching geometry students to think was difficult for all teachers. So I developed a distinctly different approach to geometry in the fall of 1987 with my fifth-period geometry class. It was all based on one statement: If Euclid could discover geometry thousands of years ago, a classroom full of today's high school students, armed with today's technology, could do the same.

Probably one of the most significant aspects of the experiment was the change in responsibility for learning. For decades the responsibility has been placed on the teacher, as if the student had no control over it. In this class, the responsibility was placed where it belonged—on the student. When I returned the responsibility for their learning to them, equity became a realistic possibility in the classroom. Each student, regardless of background or individual differences, was given the power to make this geometry class into the kind of learning environment that would be most beneficial.

At the start of the first day, I enthusiastically explained the idea to the students, encouraged their participation, and expressed my belief in them. Then I wrote the three "Given Facts" on the board—the only geometric facts I would impart the entire year:

1. Parallel lines never meet.
2. A triangle contains 180 degrees.
3. A circle contains 360 degrees.

From that point forward the students were on their own. They got to discover geometry in their own unique terms.

Every day the class worked in groups of four. These groups provided needed support in this foreign environment. Group discussions were used to investigate and experiment, and discussions involving the entire class determined the class definitions and theories. Although my support was always present, I gave no input as to the accuracy of their definitions or theories, and there was no textbook to check. In fact, the students created their own geometry book, different from regular textbooks and unique to their specific class.

The class, called Build-A-Book (BAB) Geometry, is now in its fourth year. It requires responsibility, involvement, interaction, creativity, critical thinking, and discovery on the part of the students.

Previous mathematical knowledge is not nearly as beneficial in this sort of class as in traditional geometry classes. All students—male and female, successful and unsuccessful, freshman and senior, knowledgeable and unknowledgeable, Hispanic, African American, Asian, Native American, and white—band together without anyone's having an advantage.

Build-A-Book Geometry was not created to solve the problem of classroom equity. This approach to geometry is simply a method of reaching more students than the textbook we had been using had been able to do. It is clear to me now that what takes place in BAB makes geometry accessible to all the students who find themselves in the class.

All members of the class find their own niche for success in geometry. The students learn who is skilled in which areas and aren't afraid to seek advice from each other.

> Mr. Healy, who in here knows how to do algebra real good? Lilia and I need a formula for this thing we've built.
>
> —Danny

BAB students validate each other's learning. Whether that learning involves writing, art, listening, critical evaluation, algebra, courage, or an unusual approach to the world, each member of the class has value. No one leaves the BAB class in June without having felt validated for his or her contribution. They are forever changed by the experiences of BAB, and their comments at the conclusion of the year indicate that they leave with a more positive attitude toward mathematics, each other, and themselves.

I have taught Build-A-Book Geometry for four years. In June, the response to the idea is almost uniformly positive, regardless of grade, but in September, when the class feels in new, unfamiliar territory, not all responses to BAB are positive.

> It's the first day of school in my fifth-period geometry class and the teacher dude starts out, "I'm Mr. Healy, and this class is geometry. This year in geometry. . . ." Right then I interrupted him, saying, "We aren't going to have books."
>
> It was great. I couldn't wait to see what he'd do. It's the first day of class so he doesn't even know my name, so what's he going to do?

I'll tell you what he does. He agrees with me and says, "Okay, we're not going to have books."

Man, who does this guy think he is? You can't run a class like that. How's he going to get away with this? Mr. Healy's a teacher. He can't just throw things out if he doesn't like them.

—Larry

BAB Format

Build-A-Book received mixed reviews at first. Not every student agreed with this teaching method. I didn't stuff them full of geometry, nor did I give them any input in the area of geometry. My role as the teacher changed radically. I was no longer the fountain of knowledge—I became a facilitator and a cheerleader. The students needed the motivation and support because they had a great deal of doubt about their ability to construct a subject about which they knew little.

At the start of each day, I took a few minutes to express my belief in them and talk about or demonstrate what wonderful things they had discovered on previous days. This was followed by the distribution of the investigation sheets (blank sheets of lined paper with a statement written at the top). It was their job to think and talk about the statement and write down any thoughts they came up with about the statement. While they did this, I facilitated their learning by providing materials.

Each class takes its own path in discovering geometry. To get there, many students are aided by the Geometric Supposer software (which is sometimes their introduction to the geometric vocabulary used by the rest of the world). It's a program that helps the students with many of their conjectures. For example, one group was determined to figure out how many degrees there are in any polygon. They used the computer to measure the number of degrees in each angle of different polygons they constructed and went from there to develop their theory. They relied on the Geometric Supposer to construct the figures and make the measurements, and they took that information and generalized their discovery. (This particular "discovered truth" of the student originally read as follows: "The formula for finding out the total number of degrees in any polygon is [the number of sides minus 2] times 180. This works for all polygons except for ones with 157 sides." This stood from November until March, when it was determined to work for a 157-sided polygon.)

Items such as waxed paper, mirrors, compasses, basketballs, string, wire, and beakers have all become part of the class as the students try to build their knowledge. Many of the items are purchased by the students and brought into class. They will go to a great deal of work to prove their idea. Most of these processes of discovery do not end up in the book—only the statement and a short explanation do—but the thrill of discovery is a daily occurrence.

> This class is the most interesting thing that has ever happened to me in my school life. For the first time, I can test my brain to see if it's good at logic and thinking, since in my other classes I just memorize things.
>
> —Edmund

Everyone in the class was free to test his or her brain, creativity, and ability to interact. The first day's investigation sheets contained one of the three facts I had given them. The statements for the second day's investigation sheets came from comments on the first day's investigation sheets.

The format continued in this manner for most of the year. There were better times and worse, as in all classes, but the students were almost always involved in critically thinking about geometric topics. Giving the responsibility to the students for their own learning and their own information is an overwhelming task and something none of them ever expected. They needed to use all their knowledge, creativity, and thinking skills. In addition, they had to communicate their ideas and interact with each other to develop the class.

> This is crazy. How does he expect us to come up with geometry? I need this class for my SAT's. I admit the whole idea of building geometry from the bottom up is a fascinating one. And it might be possible, if the students in the class were all A students. But the way it is, I really doubt it. There are all different kinds of students in that class. The only things we have in common are the class we registered for and a passing grade in algebra. The people in the class are all different grades and they have such different personalities. It might have been an okay idea with an honors class, but I don't think we have very many modern-day Euclids in this class.
>
> —Anna

Some students had a more difficult time adjusting than others. My most important influence was keeping their spirits positive. I provided them with the equipment and support that allowed them to investigate without the fear of failure. Although failures occurred and wrong turns were taken, it was most important that groups and individuals were never stopped by a wrong conjecture. Mathematics is full of failures, but that is not a weakness of the subject. Rather, it is a strength.

There were some students who adjusted to the idea more quickly than others. One senior boy thrived in the atmosphere of no rules, no right or wrong, and no limits on the powers of questioning. Less than two weeks after the first day, he was beginning to question the validity of one of the three facts I had given on the first day.

> Mr. Healy, you were wrong. Parallel lines *do* meet and I can prove it. [At this point Bernie brought out a basketball and some masking tape.] First, you take two parallel lines. I'll use these two pieces of tape to represent the parallel lines. You hold this end of both of these and I'll stick the ends right next to each other on the ball. [This created two parallel tangents to the basketball.] As you can see they are definitely parallel. Now you keep holding this line and I'll smooth the

> other one down on the ball. Now give me the other piece and I'll do the same thing with it. Now turn the ball over and you'll see the parallel lines do meet. Pretty neat, huh?
>
> That's not all. A friend of mine said he saw a guy on a TV show [Stephen Hawking] that said that the universe is curved. Now if that's true, Mr. Healy, then I think that means parallel lines must always meet.
>
> —Bernie

Two weeks into the year and already one of the more creative, free thinkers had challenged one of the facts I gave the class on the first day. He investigated a fact we never considered in a normal geometry class. It would be much too confusing. When he presented his idea to the rest of the class, they listened courteously and asked questions nonstop for fifteen minutes, and Bernie handled each one. One student came up and showed Bernie that there could be parallel lines on a ball that didn't meet [latitudinal lines]. Bernie listened and then demonstrated that parallel lines on a basketball didn't *want* to be parallel. According to Bernie, the outside edges of tape lines that "looked" parallel on a sphere would crumple up. He said the crumpled edges proved that the parallel lines didn't want to go that way.

An interesting demonstration—but the rest of the class was not convinced that it could really happen in a geometry that doesn't use tape for lines. So Bernie's idea never became a part of the facts in their growing book. However, from that day forward his participation in the class remained high. Bernie's own special outlook on things had not made him an outcast, as it might have in a normal geometry class, but had allowed him equal standing by using his own unique combination of talents.

The participation of some students was minimal until they finally found a topic that for some reason was important to them. The class then became a part of them, and they began to take ownership of the class.

For homework, the students develop definitions for words that have been questioned by an individual or a group (for example: Is a line thick?). On the day following a homework assignment, one group's investigation sheet calls for a review of all the homework papers to determine a class definition of each word. That definition is placed on the chalkboard the following Monday, and the class discusses it and decides on a *final* class definition of the word. That definition is typed into the computer and becomes part of their book.

A class discussion of the definition of *distance* provoked an exceptionally quiet girl to express her belief to the class.

> I couldn't let them [the class] say that "distance" was the space between two points, when I knew that there could be more than two points. I don't like to disagree, and I certainly don't like to get up in front of class, but this was different. They were confused, and I had to straighten them out, otherwise our book would have a definition in it that I knew was wrong. So I raised my hand and went up to the board to show them that it should read "the space between two *or more*

points." I knew I was right, but I expected some arguments—we argue over almost everything. But after I showed them what I thought, they all agreed and my definition became part of our book.

—Chris

Chris became a part of the class that day, and the class became part of her. Although the question of equity in the classroom is difficult to address, it is not impossible. Equity must be approached with an appreciation for the individual differences between students as well as a belief in the individual, a genuine enthusiasm for all students, and an ever-present support for each.

There are times when the students reach a definition or a conclusion that is erroneous or misleading (for example, "distance is the space between two or more points" or "$(n-2)180$ = degrees in a polygon except when $n = 157$"). I believe the students should be left on their own to correct any errors. I don't point out an error, but I do put the question on a test in the hope that it might encourage them to reconsider, though it seldom does. A teacher in Whittier, California, who is trying BAB points out blatant errors. There is no single correct way to teach Build-A-Book. It is important to me to give the total responsibility to the class to determine their own destiny. I believe that my interference would hinder that process. The errors stated earlier in this paragraph were both discovered and rectified by June.

Most of the students support each other within the Build-A-Book classroom. My job is to create an atmosphere that fosters that support. I am able to do that through my talks at the beginning of each day and my discussions with individual groups. Of course, some students are more involved than others, but I think each has a special feeling for the BAB class. When I got a letter from Chris, I felt sure that what we were doing was working.

Mr. Healy,

You know (probably) by now that I don't live here in El Monte anymore 'cause I ran away from my house. Hopefully in the future I'll explain to you why.

Right now though I want to congratulate you in your achievements in our geometry class. You're really creating some miracles in there. Keep up the good work.

I'll miss you.

—Chris

Chris stayed with her grandparents for three months in Mexico. When she returned to the city, the first place she went was the Build-A-Book class.

The day she returned, the groups functioned normally. This gave a sense of stability to her life, which had been turned upside down. Although there is no such thing as a typical day in BAB, Chris was immediately accepted back into her group and began an investigation in the fourth dimension.

A Day in BAB

The topics for the eight groups were as follows:

Groups 1, 3, and 6 were involved in similar investigations. Although all three groups had previously been investigating different things, all three groups had discovered different methods for finding the center of a triangle. They prepared their presentations for a panel presentation to the class. The class would then decide which method was correct.

Groups 4 and 5 had asked if they could combine to continue the discussion they had initiated the day before about the fourth dimension.

Group 2 discussed this statement: "A line has to be straight and has starting and ending points."

Group 7 discussed this question: "If a monogon is a shape with only one angle, is an angle a monogon? If not, is a teardrop the only monogon?"

Group 8 found class definitions for *flat, horizontal,* and *degree*. They had all the individual homework assignments, each with a definition for the three words. Group 8 had to sort through all the papers and decide on one definition for each word. On the next Monday, those definitions would be placed on the chalkboard, and the entire class would discuss them until a decision was reached. The final definition would become part of the book.

After an animated discussion that took place in many areas of the room—including the chalkboard, the floor, and the desks—Groups 4 and 5 decided that the fourth dimension was imagination and that the other groups should investigate it further.

Group 2 didn't work out very well. There were only three people in the group that day, and all three thought a line was straight only if it was indicated by the word *straight* before *line*. There was also little mention of beginning and ending points, but there were several pictures on the back of the investigation sheet. They felt this investigation was dumb.

Group 7 got very involved with the monogon and asked to continue their investigation the next day. They reported very little, and I didn't have time to talk to them. The only thing I picked up was the enthusiasm.

Group 8 did the definitions as fast as they could and then spent the rest of the time just talking about miscellaneous stuff. (Note: The definition of *horizontal* has never been very accurate. Group 8's definition of *flat* was rejected the next Monday, but the class definition was determined that day. Their definition of *degree* took weeks to settle on. I spoke with the entire class about the importance of consolidating homework definitions.)

The presentations by the three groups working on finding the center of a triangle resulted in the class's fashioning a compromise on two of them. The third presentation was not well prepared, but it sparked an investigation of imploding fractals (a phrase coined by former BAB students involved in a program last summer called Beyond BAB) that continued for two months.

Bottom Line

There is an undeniably unique brand of equity in this type of class, but the question remained: How much geometry was learned? I found two answers.

> It wasn't all that special. I'm sure I wasn't the first one to think of it. But according to people who know, my proof of the Pythagorean theorem is an original. There's over 300 different proofs. So what's one more? But the best part was how interested the Build-A-Book class was. When I presented it to them, they all listened to every word I said. Some of them even told me that this was the first time they ever understood the Pythagorean theorem. And that did make me feel good.
>
> —Edmund

It was just as I had imagined it would be—an entire class of students straining to understand the Pythagorean theorem as it was presented to them. It was like a dream. Only in the dream, I was presenting the proof, not one of my students. But I realized that the class had reached the point where they were proud of each thing they discovered and wanted to share it with the others. It didn't matter who presented and who listened. Or who questioned and who answered. People respected each other and believed in the skills of each individual.

The other indication came from the results of comprehensive tests. In late March all the geometry classes at my school were given a forty-question test from old SATs, and the scores from both approaches were compared. The BAB class average was 5 percent below the average of the textbook-taught classes. In June the highest score in the school on the comprehensive final came from the BAB class, but so did some of the lowest scores. The average score in the Build-A-Book class remained 5 percent below the average of the other classes. The following year, the BAB classes averaged 5 percent above the textbook classes on the same two tests. By this traditional measure of success, the BAB students are holding their own.

I also tested BAB students, and they might have said that "equity in the geometry class" meant that the tests I gave them were equally impossible for anyone to pass. The tests were taken from their ever-changing book. The tests questioned every part of their book, and their answers had to be verified by their book to be correct. So if there was information in their book that they disagreed with or was not clear, they would not answer the questions correctly. It was a hard lesson for them to learn, but it resulted in their reevaluating parts of their book. Reevaluating, interacting, speaking out, listening, caring, compromising, and mathematical knowledge were all part of the BAB class.

It is not my purpose to encourage others to forgo geometry textbooks. In fact, I think the quality of geometry textbooks is rapidly improving. But BAB is an alternative that deserves consideration because it fits the teaching style of many and benefits most students. As with any approach, variations make the methodology unique to each teacher and class.

During the 1990–91 school year, a former colleague used the BAB approach with his own variations. He began with Euclid's original twelve axioms. He doesn't believe in the strict hands-off approach that I use, but his class was excited about "discovering" geometry on their own.

The mathematics we teach is important, but I recognize that thinking skills and communication are more important for survival in the world. I think that people teaching these skills cannot be confined by the curriculum. They must be willing to give up control and to be positive and enthusiastic about everything the students discover. Asking questions, showing interest, giving positive reinforcement, and teaching students to believe in the value of what they discover are essential components of this approach. Those teachers interested in this method will find the students' reaction overwhelming, from their initial negative comments to their eventual belief that what they have created is as good as or better than the geometry in any other geometry books.

I feel very strongly that in our mathematics classes we must value communication, thinking, and knowledge, or we might as well not be teaching. As important as I believe mathematics to be, I am aware it is not the most important skill in life. I believe that to learn mathematics or any other subject, students have to believe in themselves and be in an atmosphere in which they can thrive. It is up to the teacher to create that atmosphere. Within each of us there exists an optimum teaching approach, and for the good of all our students, it is our responsibility to locate that approach and provide it for the students. For me, that approach requires much freedom to discover and discuss, but for other teachers it may be entirely different. I know that each teacher must seriously look for his or her optimum teaching environment. If we are not teaching in that environment, we are not treating ourselves fairly. If we don't treat our ourselves fairly, how can we treat our students equitably? And if we aren't treating our students equitably, then they are not in their optimum learning environment.

> I didn't like this class when it started and I hated math. In fact, I was pretty pessimistic about a lot of things. I don't think I knew it then, but I know it was true. I think I changed a lot. This class changed a lot of things. We all learned to believe in ourselves and in each other. I don't hate math anymore, but I still don't really love it. And, you know, now I think I can handle just about anything in my life.
>
> —Carmen

True equality in the classroom is an impossible goal because the students bring with them different skills and knowledge. Because of cultural heritage, gender, home environment, skills, and intelligence, no two students will be equal. It is important for each teacher to recognize that those differences do not have to change how we relate to the students. Although equality may not be achievable, or even desirable, providing equity for all students is a must and lies well within our reach.